SPACE SHOTS

AN ALBUM OF THE UNIVERSE

Fred Hapgood

Introduction by Michael Collins
Apollo 11 Astronaut

Times
BOOKS

Published by TIMES BOOKS, a division of
Quadrangle/ The New York Times Book
Co., Inc. Three Park Avenue, New York,
New York 10016

Published simultaneously in Canada by
Fitzhenry & Whiteside, Ltd., Toronto.

Printed in U.S.A.

ISBN 0-8129-0823-6

Library of Congress Catalog Card
Number: 78-68713

Created and produced by
Tree Communications, Inc.
250 Park Avenue South
New York, New York 10003

CONTENTS

INTRODUCTION

Halley's comet last hove into view in 1910 and will appear again in 1986. Although the comet will not have changed much in terms of a human life-span, the solar system the comet frequents has changed in a very substantial way during that interval. In 1910, inhabitants of our sun's third planet could propel themselves at forty mph; by 1968 we were streaking away from our home at "escape velocity" – 17,000 mph – with the modest goal of reaching the moon, our planet's only natural satellite. By 1986 we will have dispatched our machines out of this solar system and could certainly – if we willed – visit other planets in person. Just as the human speed limit has increased geometrically since Halley's comet's last visit, so has our knowledge of our surroundings. I believe that we are now seeing greater breakthroughs in the field of astronomy than in any other science, and those of us who consider this the Age of Astronomy rejoice that this is so. On the other hand, the true believer could also call this the Age of Frustration – because so few of us have been fortunate enough to have ventured above our planet's atmosphere. In an important way, this book serves both audiences – those who rejoice at the new knowledge bombarding us daily, and those who chafe because they cannot go there themselves.

Personally, I have no craving to venture near a "black hole" or even to get much closer to our sun than Acapulco in July, but I would like to hike on Mars and take a dive into the thick atmosphere of Titan, Saturn's intriguing satellite. My own brief trip away from Earth in 1969, orbiting the moon while Neil Armstrong and Edwin Aldrin walked on it, certainly whetted my appetite for venturing farther out. I liked what I saw and wanted more of it. When I was on the back side of the moon, over two thousand miles away from my colleagues and nearly a quarter of a million miles from the rest of you, in total darkness, with the Earth nowhere in sight, I felt I *belonged* there, to wander silently among the stars. There being no atmosphere to spoil my view, the stars gleamed with a steady brilliance – alluring, inviting – welcoming me out into their domain. I get the same feeling again as I turn the pages of this book. Certainly old friends are here (the glorious Earth and seemingly lifeless hulk of its moon), but a host of fascinating strangers appear as well.

Michael Collins, Apollo 11 Astronaut

Opposite: A bubble of interstellar gas glows with tremendous energy as it expands through space, thousands of times larger than our solar system. It has been suggested that some of these clouds, called nebulae, carry the basic ingredients of life. (Hale Observatories)

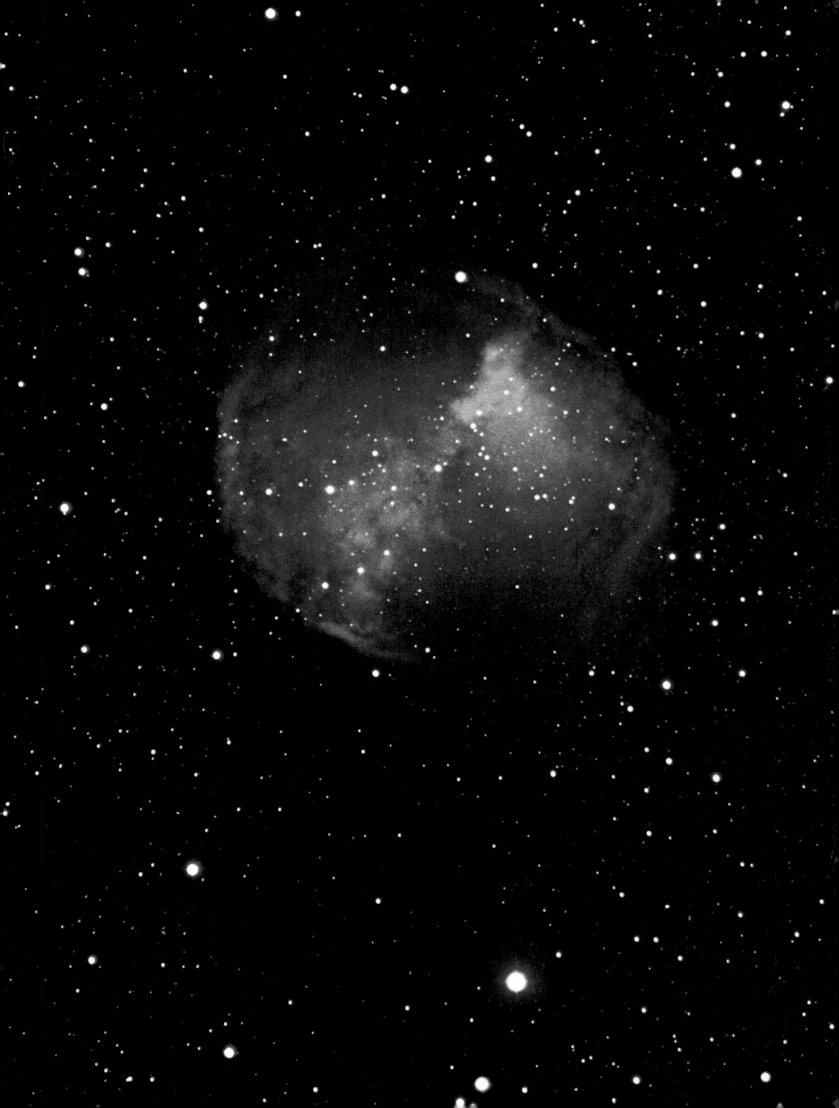

THE UNIVERSE

When you look up into a clear night sky, with its stars blazing from one horizon to the other, you see the universe. But what is that? Is it a single thing, like a fish or a molecule of water, that acquired its properties and became itself all at once? Or did the universe drift into existence like a sand dune, an accumulation of parts? Does it matter to all the galaxies that they find themselves together in the same universe? Do they affect each other like people jostling about in a crowded room, or does the universe impose a common discipline, a controlling force, that marches all its members along the same path? And however it began or however it conducts the rest of its cycle, how will the universe end: as a whole, dying as we do, losing its properties and faculties all at once? Or will the universe slide slowly out of existence (whatever *that* might mean) with each galaxy pursuing its own independent fate?

These are daunting questions, and yet they are important enough so that every civilization known in our history has anwered them. In general the answers have been that the universe, as a whole, is very old, cyclic, and profoundly interconnected.

"I have known the dreadful dissolution of the universe" (Vishnu is quoted in the Hindu *Puranas*). "I have seen all perish, again and again, at the end of the cycle. At that terrible time, every single atom dissolves into the primal, pure water of eternity, whence originally all arose. Everything then goes back to the fathomless, wild infinity of the ocean, which is covered with utter darkness and is empty of every sign of animate being."

Another of the world's great cosmologies was formulated by the Buddhists. They agreed that the universe was old and cyclic and thought that a memory of these cycles was a sign of spirituality. According to Stanley Jaki, a student of these matters, the lowest gods could only retain some recollection of forty of these cycles, the next grade up remembered 1000, the next 100,000, and so on, until one reached the higher Buddhas who had access to millions of cycles. Many of these calculations yield an estimate of the age of the universe that is strikingly close to that given it by contemporary astronomers.

Most of these ancient civilizations not only agreed that the universe was big and old, but that its nature as a whole dominated its parts. The Taoists believed that all of nature was a product of a continual universal interplay of opposites. The Egyptians believed that the universe had a single, all-pervading rhythm and that it was the duty of man to discover this rhythm and identify himself with it. Halfway around the globe the Aztecs had ideas that seemed very similiar although their understanding of the natural cycles was more violent and cruel.

Of all these great cultures our own stands out as caring less than any other about these great questions. We observe the parts – the planets, the stars, the galaxies – for their own sake, regardless of what they say or don't say about the final question of existence. Ours is a very strange civilization, and some believe it evolved because Euclid and Abraham happened to be born in the same part of the world. Judaism believed that the world was made by a rational lawgiver who had also made man in his own image; Euclid showed that it was possible to discover laws that were true forever. Greek and Judaic elements may have fused to give us the conviction that natural, permanent laws (ones that we could discover with our own talents) were out there waiting to be found.

For a great many centuries we thought we saw a universe very different from those glimpsed by other civilizations: a universe created a fairly short time ago and certain to end once and for all (no cycles here) in a Day of Judgment. But over the last four hundred years, as we followed the trail of one discovery after another, farther and farther outward, our own views and those of the ancient civilizations appear to have converged. Many scientists now believe that a cyclic universe is a distinct possibility; and everyone agrees that the cosmos is much older and larger than medieval astronomy allowed. Einstein showed that the expansion of the universe affects all its parts. And many believe that the universe as a whole will one day collapse together again into "utter darkness" – a black hole.

What is bringing us to this common realization are the objects pictured in this book: the bodies in the heavens. Our tradition is such that we move away from our sphere stepping-stone by stepping-stone: the planets of our system as they orbit around the sun; the sun, as it and 100 billion other stars orbit together around the center of our galaxy; the twenty-one galaxies of our own cluster of galaxies; and the billions of other galaxies strewn across space, as many in the universe as there are stars in the Milky Way.

Like any other civilization we are a product of our environment, and increasingly, as we learn to control the Earth, it is the universe that is our environment. It affects and molds us as we reach farther out into it. The direction our changes take seems to be moving us toward the perspectives held by most of the great civilizations in the history of the planet.

(Editor's note: 1 billion = 1,000,000,000)

EARTH

Five billion years ago there was no Earth, or sun, or solar system at all — just a cloud of dust and gas in this region that was beginning to collapse. As the cloud fell together, enough matter collected in one place to form the sun, and the gravitational forces of the new sun flattened out the cloud around it into a thick disk. Small particles whirling in this disk gathered others and began to grow by accumulation. What became the Earth, orbiting 93 million miles away from the sun, had cleared its area in the solar system and reached its present size about 4.6 billion years ago.

Shortly after it was formed (or perhaps while the process was ending) huge bodies of planet-building debris called planetesimals sometimes hundreds of miles across slammed into Earth, heating the already hot planet still more. Earth literally melted under the force of its own weight pressing in, the impacts, and radioactive decay and became chemically differentiated as the heavier elements like iron and nickel sank to the center and lighter elements like silicon rose to the surface. Gases driven from the churning crust formed a primitive atmosphere composed mostly of methane, ammonia, and water. As the planet cooled, water condensed from the atmosphere to form the oceans. Ultraviolet radiation from the sun reacted with the atmosphere to make organic molecules that drifted down like manna into the protection of the oceans where, less than a billion years after the planet's birth, the molecules embarked on an evolutionary career that continues in every life-form today.

Earth was formed about 12 billion years after what we assume to be the beginning of the universe (the Big Bang), and Earth will cease to exist when the sun swells out in its dying phase and vaporizes our planet (we guess) about four or five billion years from now — long before the universe ends (or begins again). We are children of a moment in the universe, and when we look around us we can find every stage of our cosmic life-cycle. The farther we look the better we understand how magnificently small we are.

An extraterrestrial visitor wandering into our solar system now would find Earth, turning jewel-like and vulnerable in its transparent skin, releasing tiny seeds into its own near space and toward its moon and fellow planets. Only twenty years after we started to travel in space, an unmanned vehicle is leaving the solar system altogether bearing a plaque explaining who we are and a recording of the sounds of twentieth-century life on Earth. Man is coming of age in the community of the universe and introducing himself to the stars.

We can no longer imagine that we are alone in the universe. Although we have never seen another solar system, it is reasonable to suppose that ours is not unique. If only a small fraction of the stars in our galaxy have planetary systems, and if only a small fraction of those would be suitable for life, the number of life-bearing planets would still be immense. Looking at the entire universe, we can suppose that it is not empty but is more apt to be teeming with life.

Opposite: While a storm swirls over the U.S. Northeast, most of the country enjoys clear weather on a summer afternoon. Earth is predominantly blue, from the oceans that cover two-thirds of its surface, and white, from the cloudcover and polar ice cap; the continental landmasses are various shades of brown and gray-green. The planet does not show its full disk because the sun is shining from the photographer's left. (NASA)

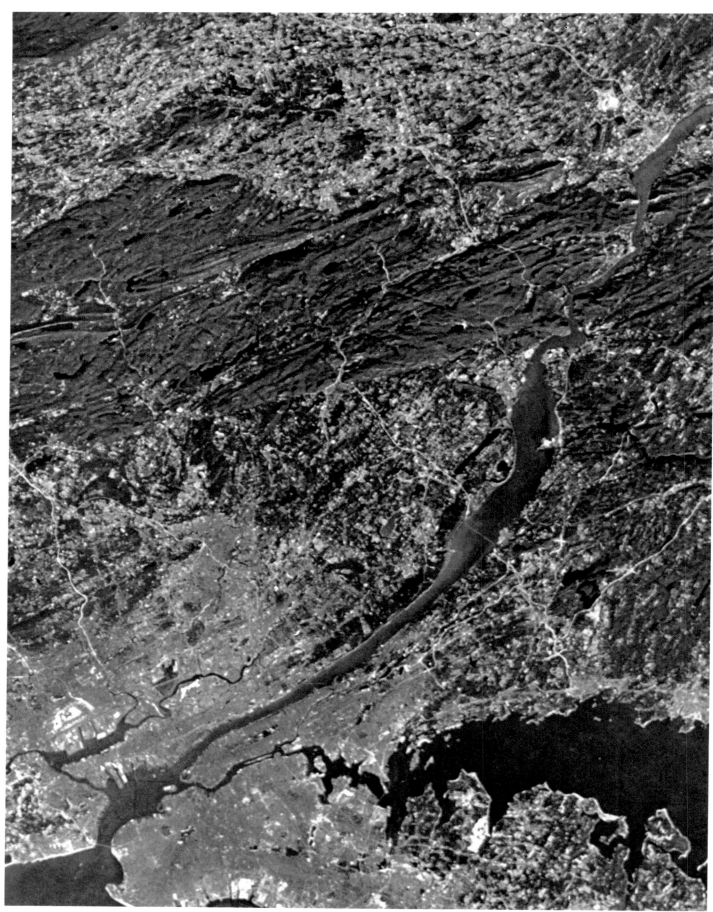

Above: Life on Earth shows in various hues in an infrared photograph of the New York City metropolitan. Unchecked vegetation in rural areas is deep rust, the more densely inhabited suburban regions are light pink, and the essentially unvegetated surface of the urban area itself is blue-gray. (NASA)

Right: Berry Island in the Bahamas and its surrounding shelf of barely submerged sand look like a giant conch shell. Ocean currents drift the glistening underwater sand like wind drifts sand in the desert. (NASA)

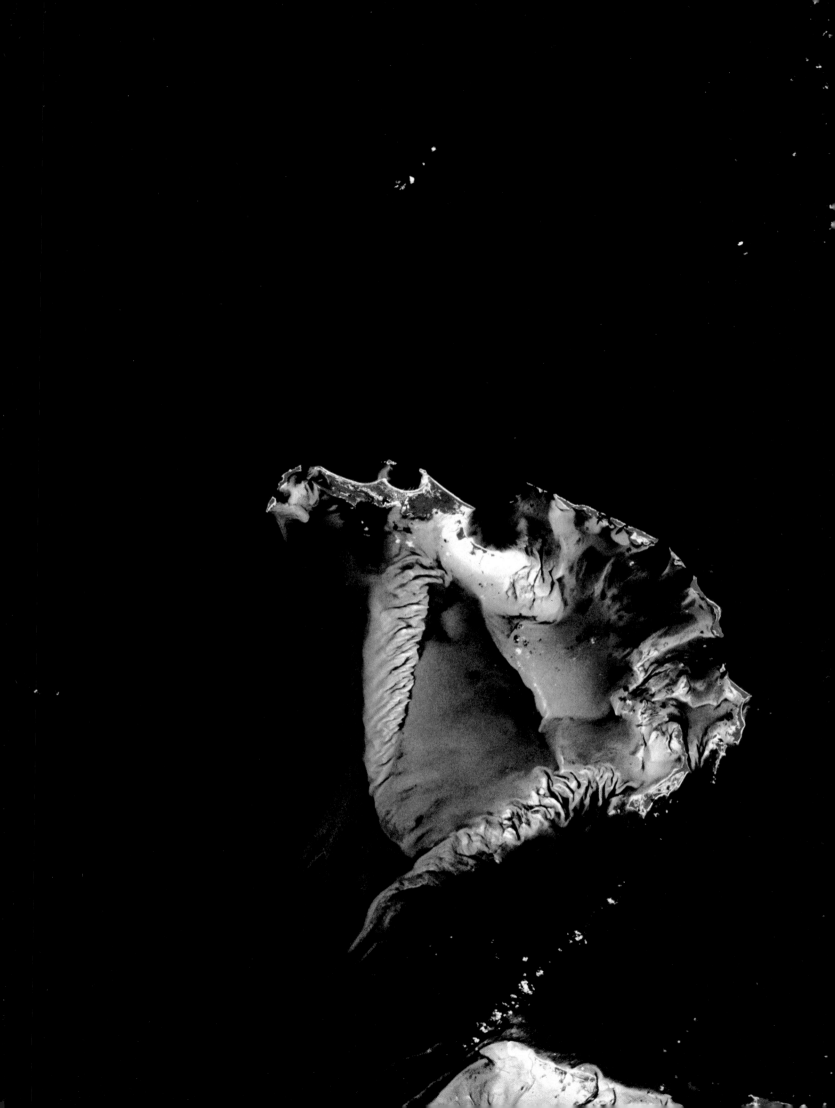

Above: Seen from directly above, a dormant, snowcapped volcanic cone in New Zealand is almost perfectly circular at its base; a smaller cone, without snowcover, rises to the right of the central peak. See pages 31 and 32 for extraterrestrial volcanoes. (NASA)

Right: Clouds spiral around the center of a tropical storm as it spins counter-clockwise over the Pacific Ocean in the northern hemisphere. The direction the storm spins is determined by the Earth's rotation: if it were in the southern hemisphere it would spin clockwise. (NASA)

THE MOON

If Earth's moon were a giant sculpture hammered out by an artisan god and left hanging in the sky, no one could doubt that the intention was to teach us about deadness. On Earth the seas are liquid and they have mothered life; on the moon they are stone. There is either killing brightness on the moon or complete darkness – no gentle transitions. Everywhere the surface shows the effects of intense pounding. There is no atmosphere whatever – no life.

Our satellite is too small to have retained sufficient heat to run an active geology for long or to have kept an atmosphere as barrier, buffer, and circulatory system to further change its face. As a result the moon is an exhibit from an early age in the solar system. Its most prominent features – easily visible to the unaided eye – are craters and basins (called seas) that record the Great Cataclysm, a period early in the development of the solar system when a blizzard of planetesimals swept through the system, smashing into the moon, Earth, Venus, Mars, and Mercury. While Earth's active geology changes its surface features in brief millions of years, the moon's geology has been frozen for nearly 4 billion years.

Sterile as it is, the moon probably had more to do with life than simply record the conditions shortly before life began. Recent thinking on evolution stresses the importance of ocean tides and their gentle, persistent rhythms. Tides, the moon's gravitational effect on the liquid portion of this planet, may have been important twice in the emergence of life on Earth.

The complicated biomolecules that are the components of all living organisms are extremely long chains that repeat the same subunits over and over again. If, as some believe, the earliest forms of these biomolecules formed on the surface of certain clays, then the most obvious way they could have been supplied with a steady source of the sub-unit molecules was through the action of tides washing over the clays endlessly while these molecules gradually lengthened.

When life had evolved, at first as animals that lived only in the oceans, it was probably the action of the tides, constantly throwing life forms onto beaches and into the air for short periods, that was responsible for the evolution of species able to live on land. The recurring phases of the moon probably taught early man a basic way to count and gave him a sense of time. Perhaps most important, in the long perspective, it gave him something to reach for.

Man journeyed to the moon for the first time at the end of 1968 when *Apollo 8* traversed the 250,000 miles from Earth and spent a few days in lunar orbit. Since then the surface has been visited six times. Altogether man has inhabited the lunar surface for 13 days, brought home 845 pounds of rocks, and left several tons of instruments, space hardware (including 3 cars), innumerable footprints, tire tracks, flags, family snapshots, and other mementos of the visits. Were man to disappear at this moment, the moon would record these few remains of his evolutionary progress as faithfully as it has preserved a record of our early solar system.

Opposite: The moon (as seen from an approaching spacecraft) is dominated by craters and large, relatively flat areas called basins or "seas", the result of vast upwellings of lava that spread across the surface during a billion year period. (NASA)

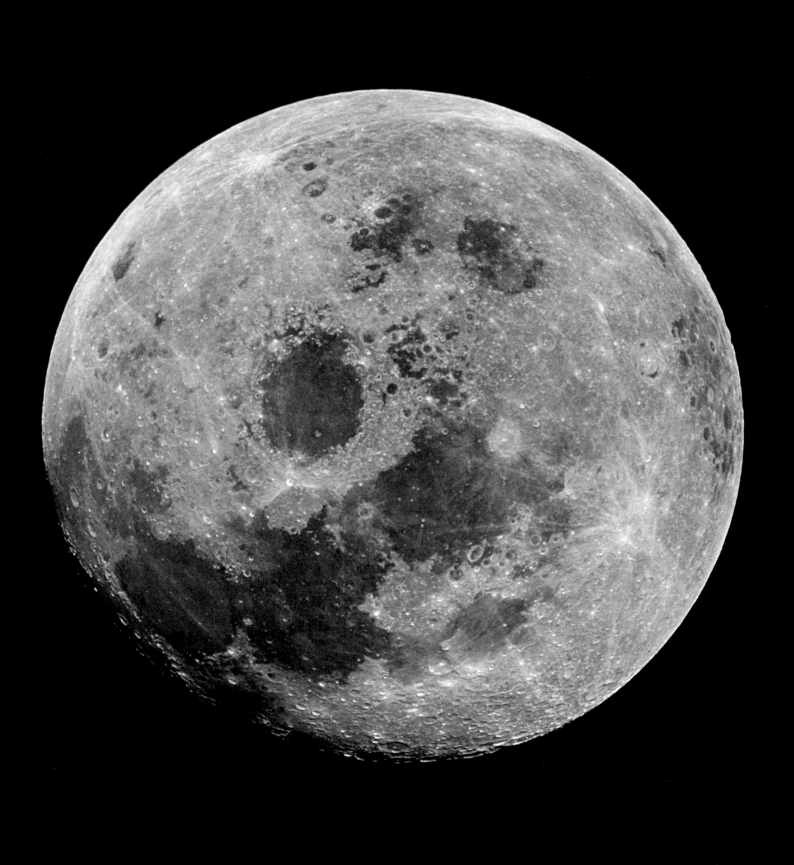

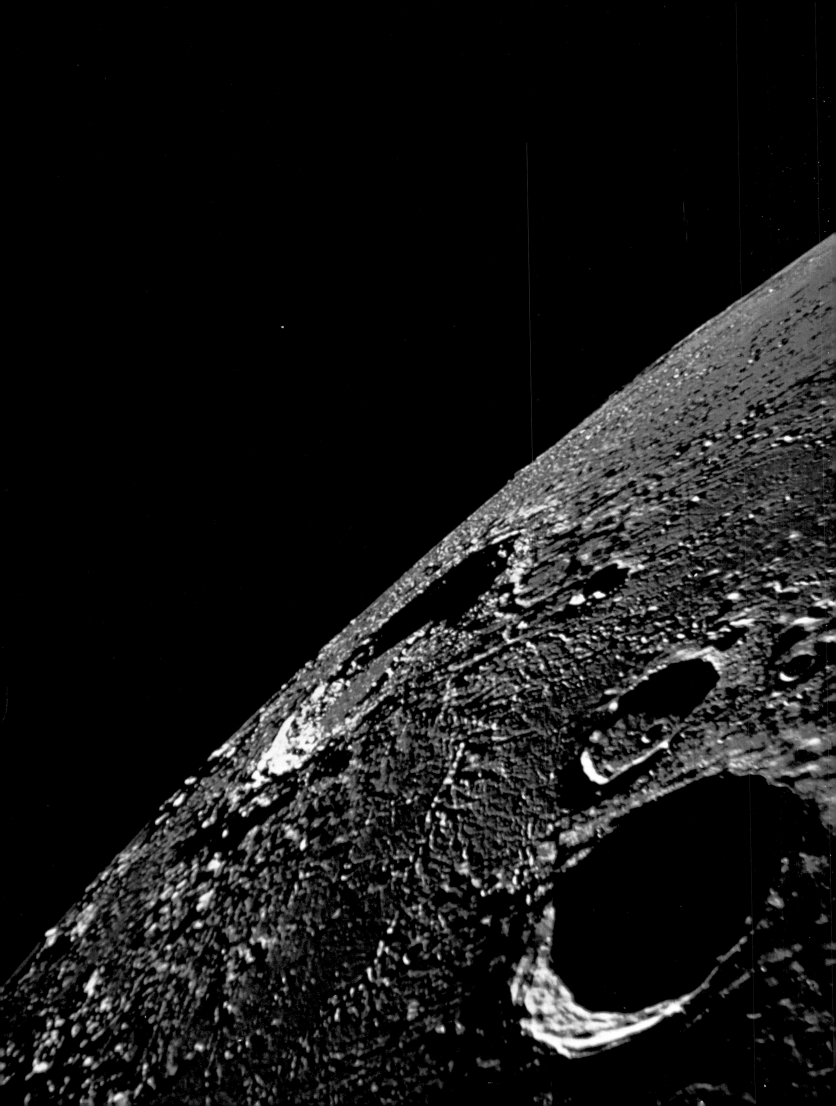

Left: From orbit around the moon the pilot of *Apollo 12* photographed the crater Copernicus while two fellow astronauts spent time on the surface. The inner walls of Copernicus have slumped in giant landslides. (NASA)

Above: Crater Tsiolkovsky is around the eastern rim of the moon and had never been seen before the first orbital explorations of the dark side. The mountain rising in the center of the crater is typical of impact structures on the moon, Mercury, and Mars. (NASA)

Following page: Man last visited the moon in 1972 when *Apollo 17*'s lunar module spent three days on the surface in the mountainous Taurrus-Littrow region. The tire tracks give an idea of the consistency of the dusty plain; the mountains in the distance are several thousand feet high. (NASA)

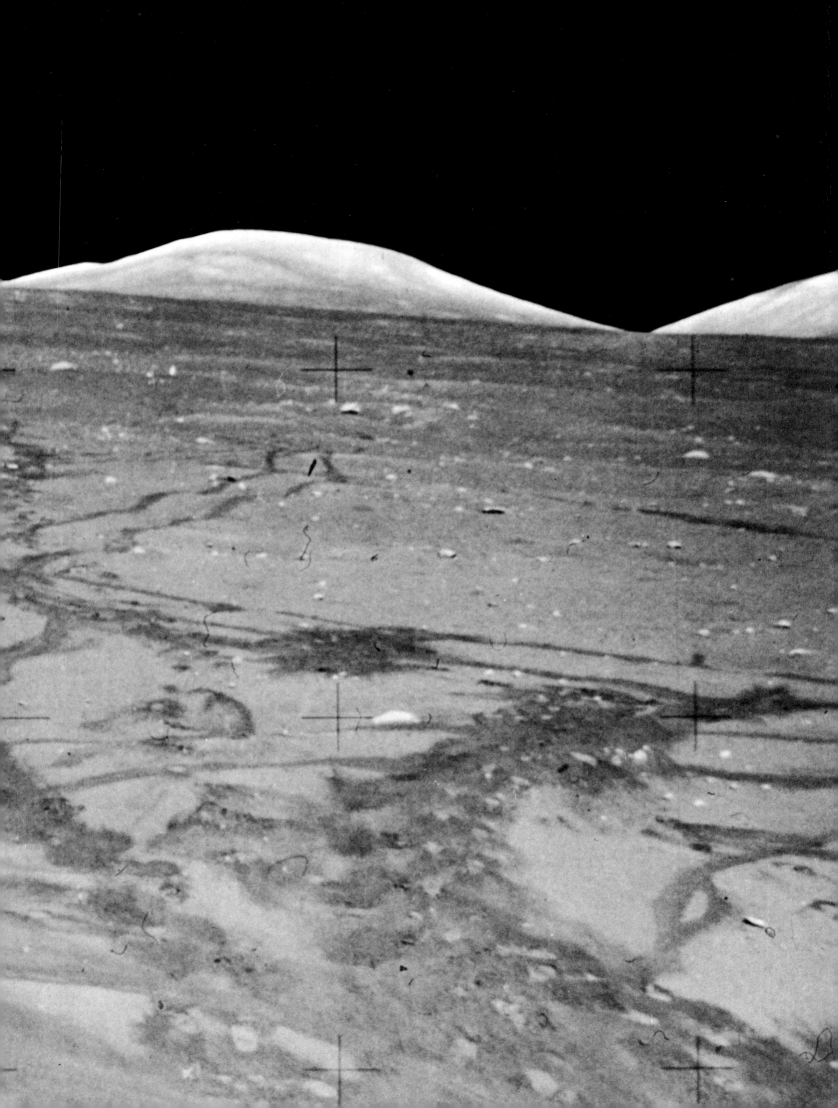

ASTEROIDS

When asteroids were first discovered in the early 1800s, they were given names such as Ceres, Pallas, Vesta, Cybele, and Psyche. Now when new ones are charted, they are called Fanny, Frigga, Adelheid, Bamberga, and Dembowska. No matter how august the family, one can feel sincere deference to only so many of its members. Ironically, one named Eros may pose some danger to Earth.

About 2,000 of these minor planets are now well known, in the sense that at any given moment a computer can tell exactly where they are. We have vague clues of another 2,000 or more whose exact orbits are still unknown. Astronomers sometimes refer to asteroids as the "vermin of the skies," since they streak through time exposures, laying down a trail through precisely the point of interest.

But this attitude has another side. It is believed that orbiting somewhere between Mars and Jupiter are the "genesis rocks" – bodies that have remained unchanged since the beginning of the solar system. Like comets, but for a different range of substances, these rocks could be the reference against which we can measure all the changes that have happened – the points that mark the beginning of the story.

Asteroids represent not a shattered planet, as was once thought, but a stillborn one, the planet Aztec, whose evolution was halted at the very beginning of its life cycle. Planets are believed to require tranquil circumstances for their birth. An embryonic planet is very small and has no gravitational field to speak of, so it can get larger only by slowly reeling in neighboring particles traveling close to it, in the same direction, at the same speed. Anything that disturbs the neighborhood so that the orbits and velocities of the local materials get mixed up would prevent a major planet from gathering itself together.

Jupiter, a giant even in its youth, seems to have played spoiler to what are now the asteroids. Its vast gravitational energies at birth so agitated the material in its vicinity that when particles met they ricocheted apart, fragmenting instead of combining in a gentle, mutual capture. What is left are the primal materials: substances that have never been fed into the dynamic cycle of planet building.

Most asteroids orbit the sun between Mars and Jupiter, but some have been perturbed into eccentric orbits that intersect Earth's. One group of asteroids called the Apollo's presently offers the best possibility of someday contributing one of its members to the scenery on Earth (although the chances are comfortably remote). The consequences of such a violent encounter would depend on the size of the asteroid smashing into the planet: one of the Apollo's smaller members – perhaps 200 meters across – could cause local devastation; a large one, like Eros – measuring about 7 by 19 by 30 kilometers – would be incredibly destructive over a considerable area and might well cause major planet-wide changes. If we find Eros headed our way, we will have about six months to get ready.

Opposite: Phobos, an asteroid-like body about 10 miles in diameter, is actually one of Mars' two tiny moons. Asteroids range in size from the largest, Ceres, with a diameter of 480 miles (and as much mass as all the other asteroids combined), down to ones only a few hundred yards across. (NASA)

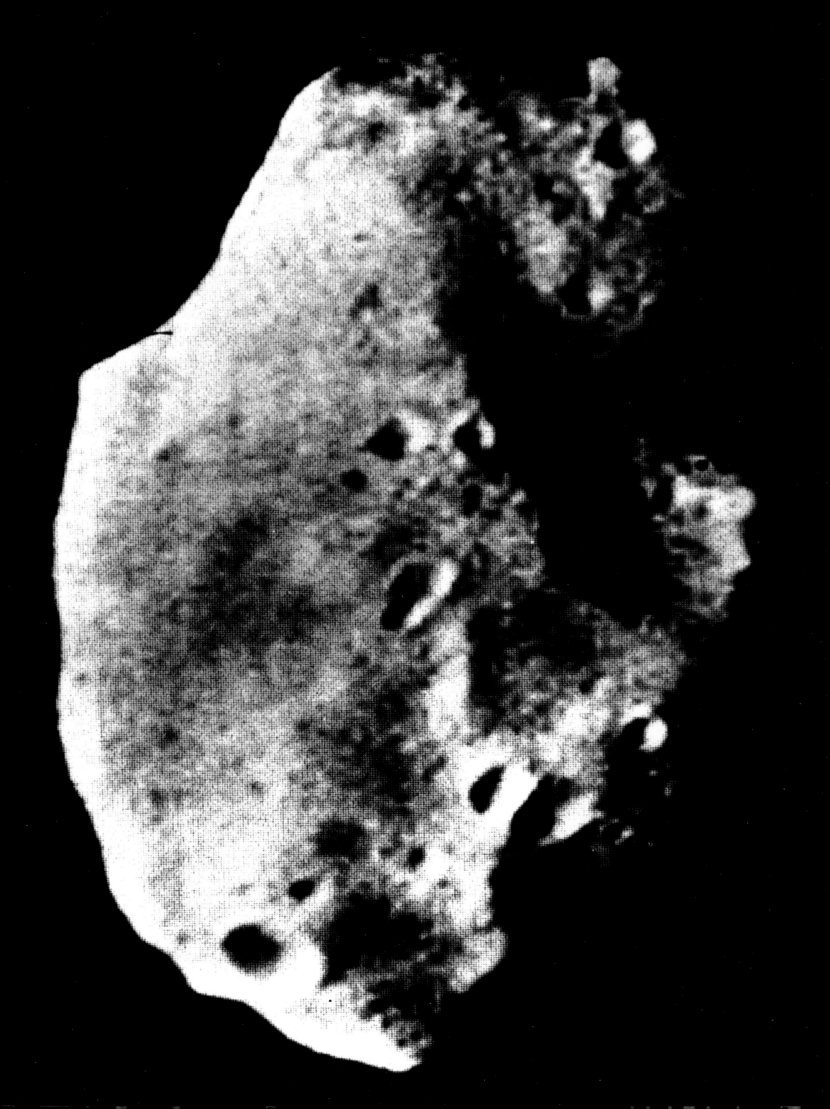

COMETS

Comets are profoundly interesting to the storytellers in our astronomical community because comets (like the asteroids) are relics of early solar history. They left our region of the system billions of years ago and have been in the cold storage of interstellar space ever since. When a comet returns (streaking around the sun) it comes from the past, a fragment of primeval stuff. Comets give us an opportunity to see what changes there have been over the last nearly 5 billion years.

The ignition of the sun drove the cometary materials out of the solar system. Before the sun grew large enough to burn, this region of the universe was made up of a large cloud of hydrogen, oxygen, nitrogen, and carbon, and some vanishingly small traces of other heavier elements. The energy the new sun produced blew the lighter elements into the outer reaches of the solar system. Hydrogen, oxygen, carbon, and nitrogen clustered together into fluffy snowballs (with something of the feel of cotton candy). Enough of this substance adhered to the larger balls to become the outer planets, and some of the smaller balls were flung to the farthest reaches of the solar system by the gravitational dynamics of the growing outer planets. They float there today – for the most part, a large, loose cloud, a Sargasso Sea of billions of comets. Occasionally the gravity of other stars disturbs this cloud, a few comets are shaken loose, and they plunge back toward the system from which they were originally ejected.

The appearance of comets has often caused great turmoil and alarm among the inhabitants of this planet. There is some irony in that, for the inhabitants of the third planet share much with comets, chemically speaking, and relatively little with Earth itself. When the inner planets were formed, they were assembled out of the elements that had not been blown away from the young sun; silicon, iron, aluminum, magnesium, and others. But we are made up of very small quantities of these elements. Organisms are made up mostly of hydrogen, oxygen, carbon, and nitrogen – the stuff of comets (and stars, and the primitive nebulae).

How did these light elements come to be here on Earth so that they were on hand to create us? There are a few theories as to how the lighter elements and molecules (water, carbon dioxide, carbon monoxide, ammonia, and others) survived the ferocious solar wind that scorched the inner planets. One theory holds that these elements and light molecules *did not* survive at all, that after the sun had announced itself so dazzlingly, Earth was left a scorched cinder. The lighter elements were *restored* to Earth, restored by comets. The same forces that threw most of the smaller comets out of the solar system would, the argument runs, have thrown some of the comets back into it, down toward the charred inner planets. Some would have entered Earth's orbit and over a period of time, been swept up by the third planet, crashed into the surface, and contributed material for molecules so complicated and ambitious that one day they would begin to search for the story of their origin.

Opposite: Comet Kohutek, photographed in ultraviolet light from *Skylab,* was expected to shine brilliantly in the night skies when it passed near Earth in 1973-74, but it was a disappointment, hardly visible to the unaided eye. Kohutek, with its tail here streaming 3 million miles behind, will not pass near Earth again for thousands of years. (NASA)

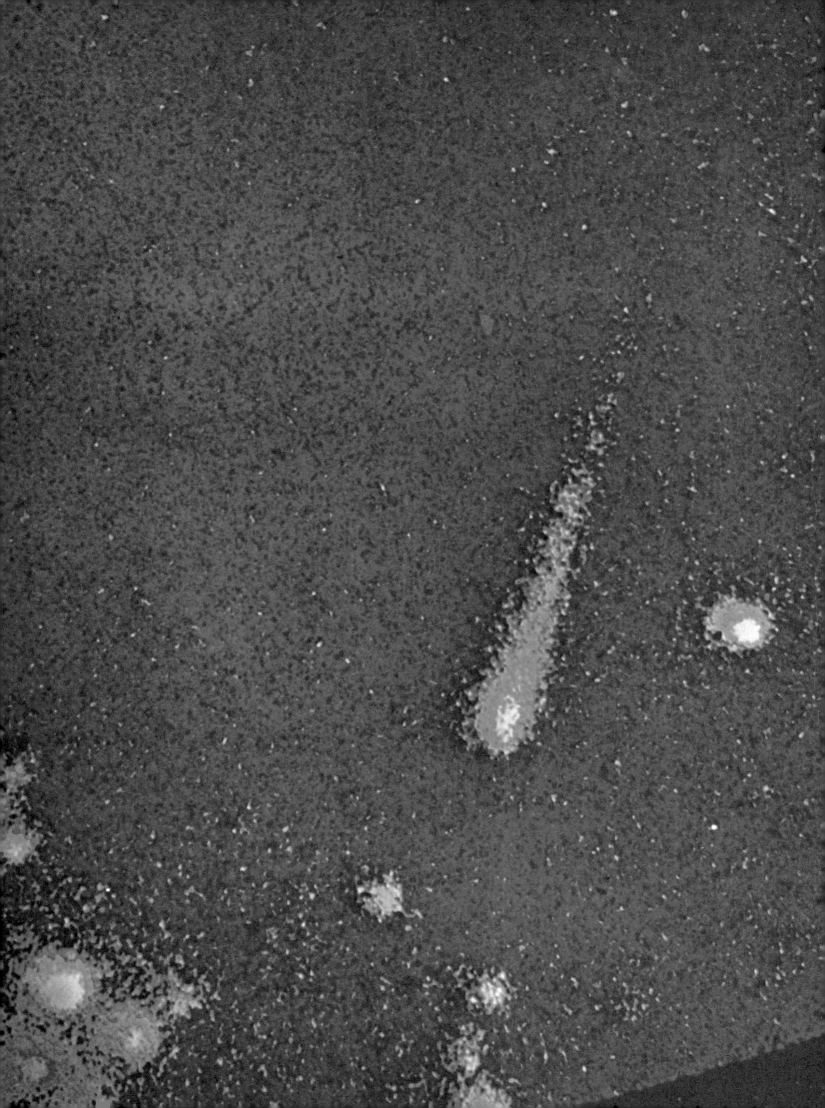

 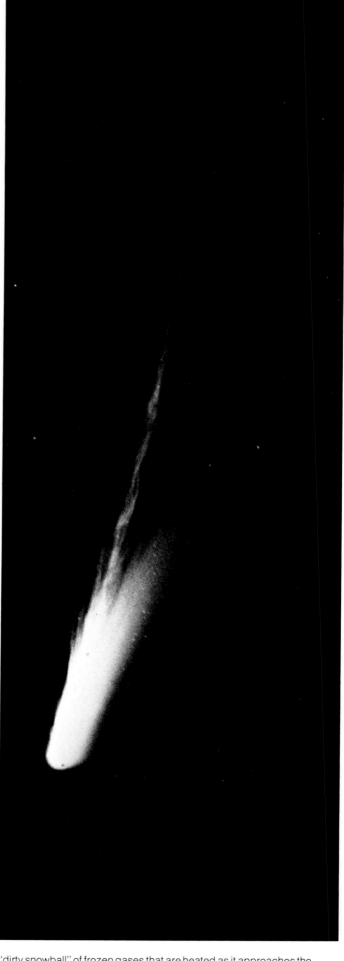

Four views of Mrkos comet, taken in 1957 on August 22, 24, 26, and 27, show changes in its tail. The head of a comet is thought to be like a "dirty snowball" of frozen gases that are heated as it approaches the sun; the solar wind drives some of the gases away from the head, creat-

ing a tail (which always points away from the sun). Changes in the appearance of the tail are due to variations in the rate gas boils away from the head and fluctuations in the intensity of the solar wind. (Hale Observatories)

VENUS

Man has long felt a deep attraction to the second planet; its very name – Venus – says this. The sun and the moon aside, Venus is the brightest object in the sky – inviting man to brood on it.

Venus is our closest planetary neighbor and, in many respects, our sister, having about the same size and density and possessing a considerable atmosphere (unlike Mars or Mercury). Since the influence of local circumstance has been thought to be as important to planets as to people, Venus has been expected to have many features in common with Earth. Qualified observers used to believe the second planet to be a better prospect for extraterrestrial life than Mars.

The picture that has emerged on closer inspection is of a body so alien that it makes Mars look homey and familiar by comparison. The Venusian atmosphere is 90 times heavier than Earth's; surface pressure on Venus is equivalent to the pressure at 3,000 feet below the surface of our oceans. This crushing "air" has virtually no oxygen; instead it is composed mostly of carbon dioxide and is thick with clouds of acids: sulphuric acid, hydrochloric acid, and hydrofluoric acid. The hot Venusian rain is probably the most corrosive fluid in the solar system. To add to the nightmarish conditions, the acid clouds are driven around the planet once every four Earth days.

Venus is a killer planet. Its surface temperature is hotter than a self-cleaning oven. Although it appears that at least one volcano and a large crater have been found by radar observation, the surface is predominantly smooth, with few prominent geological structures. The hot, corrosive atmosphere can probably bring down mountain chains like sand castles. A biology of any sort is unimaginable.

The main features of the planet are now well known, but a basic question remains: how did it develop so differently from Earth? Why, for example, are all those acids in the atmosphere? Is it because the extremely high temperatures evaporated the acids from the surface? But where does that high temperature (900°F) come from? No doubt it is caused by the enormous capacity of the massive carbon dioxide atmosphere to absorb and retain sunlight. But where does all the carbon dioxide come from? Earth seems to have just as much as Venus, but the difference is that on Earth the carbon dioxide is bound up in rocks; on Venus it is mostly in the atmosphere. What force could strip the CO_2 (carbon dioxide) out of the Venusian rocks? Is it the high surface temperatures?

The questions go around and around; so far we do not know what started the process running. We ought to know this; more is involved in this search than simple curiosity. On Earth we have been burning organic materials (wood, coal, oil) for a long time, and in doing so have released larger and larger quantities of CO_2 into our atmosphere. If, as is quite possible, Venus started its slide into hellishness by the chance occurrence of a little extra CO_2 in its atmosphere, then the warning it can give us may turn out to be one of the most important achievements of the space program.

Opposite: The swirling acid clouds of Venus show clearly in a computer-enhanced mosaic of images taken by *Mariner 10* from a distance of 450,000 miles. The planet appears blue (not its true color, yellow) because of filters used in processing the image. (NASA)

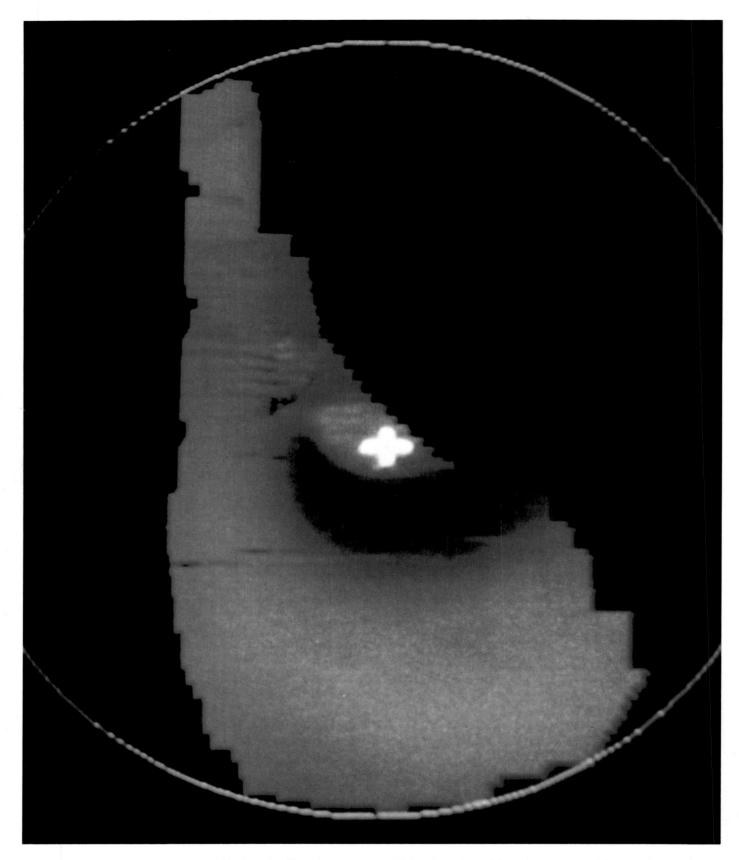

Above: A heat-sensitive camera sketched this view of the Venusian north pole (marked with a cross). The dark area around the pole is a part of the atmosphere 90 degrees warmer than its surroundings and may be one of the bands shown in the ultraviolet photograph on the preceding page. (NASA)

Right: The surface of Venus is masked by a featureless haze as the *Pioneer Venus Orbiter* approaches the day side. Venus hardly turns on its axis at all, completing a rotation only once every 225 days, but beneath the yellowish outer atmosphere thick clouds race around the planet once every four days on violent winds stirred up by the high surface temperature. The average height of surface features on Venus seems to be less than on any of the other inner-planets. (NASA)

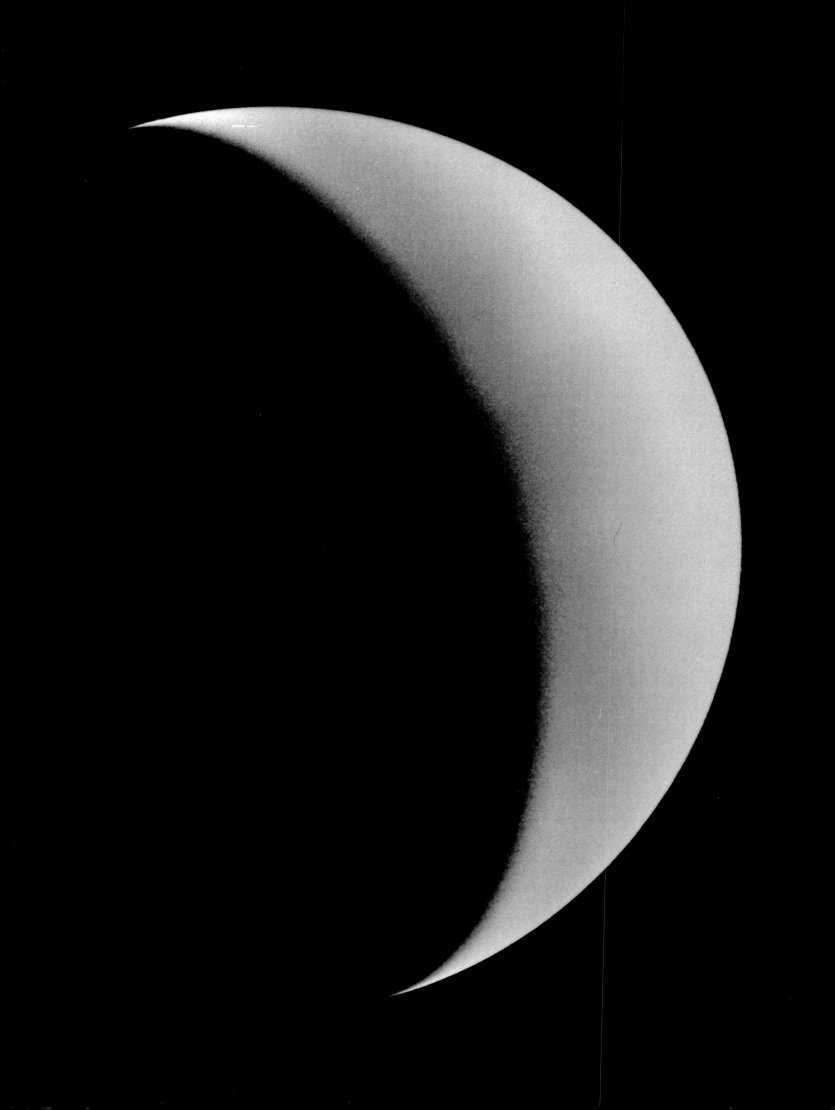

MARS

Ever since the Italian astronomer Schiaparelli released his maps of the Martian *canali* in 1877, there has been only one question of interest to the nonspecialist about the red planet: Is anyone there?

When *Viking 1* and *Viking 2* blasted off from Cape Canaveral in the summer of 1975, they carried instruments designed to radio back an answer as to Martian life. Even deciding how to pose the question was difficult because, after all, what *is* life? What exactly were they looking for?

Cameras were included, of course, to pick up gross clues like trees or tire tracks (or arid plains of broken stones, as they saw). But cameras miss a lot, so the *Viking*s carried sophisticated laboratories to perform robot experiments. In one test, a sample of Martian soil (delivered into the lab by a mechanical scoop) was mixed with terrestrial nutrients and the gases given off were analyzed. On Earth such a mix would show a gradual increase in the amount of carbon dioxide (CO_2), and other gases, given off as breathing organisms fed on the nutrients and multiplied. On Mars, when the nutrient was fed to the soil, CO_2 and some oxygen were given off in a burst. Neither terrestrial biology nor soil chemistry behaves this way. The sample was heated to the point where any biology would be destroyed, and the reaction stopped. But heat can also destroy chemical reactions: were we looking at evidence of life or just strange chemistry?

A second test introduced radioactively labeled carbon dioxide (CO_2) and carbon monoxide (CO) into a sealed chamber containing another soil sample. After five days the soil was analyzed to see whether any of the labeled gases had ended up in organic molecules. If so, that surely would be evidence that something was making organic molecules (like sugars) out of CO_2 and CO. Seven out of nine times the results showed trapped fragments of synthesized organic molecules. But when the chamber was sterilized with high temperature, the reaction continued! Mars either has a chemistry that behaves biologically or a biology that behaves chemically. Nothing like it exists on Earth.

There is no rain on Mars. The air is so thin and dry that any water released on the surface would evaporate instantly. Martian geology suggests this atmospheric thinness because the surface features shown in these photographs are sharp and crisp – not the qualities one would see on a planet where the air was thick enough to carry sand. Much of whatever erosion there is on Mars seems likely to have been caused by ultraviolet radiation from the sun, subtly breaking down the structure of the rocks.

However, there is evidence that liquid water did once exist on Mars and that the air did once carry substantial freight. Mars is rich with channels and in many cases they form an integrated and extensive drainage system. There are large deposits of water ice and wind-blown dust and fields of sand dunes at the north and south poles. All this material must have been carried by something from somewhere. The mysteries of the planet await further investigation.

Opposite: Plumes of water ice stream toward night from the top of a huge volcano on the dawn side of Mars. The middle of the planet is scored by the 3,000-mile-long rift canyon, Valles Marineris; the south pole is covered with a carbon dioxide and water ice cap. Although the atmosphere is very thin, and cloud cover virtually nonexistent, Mars is regularly covered by planet-wide dust storms that last for months. (NASA)

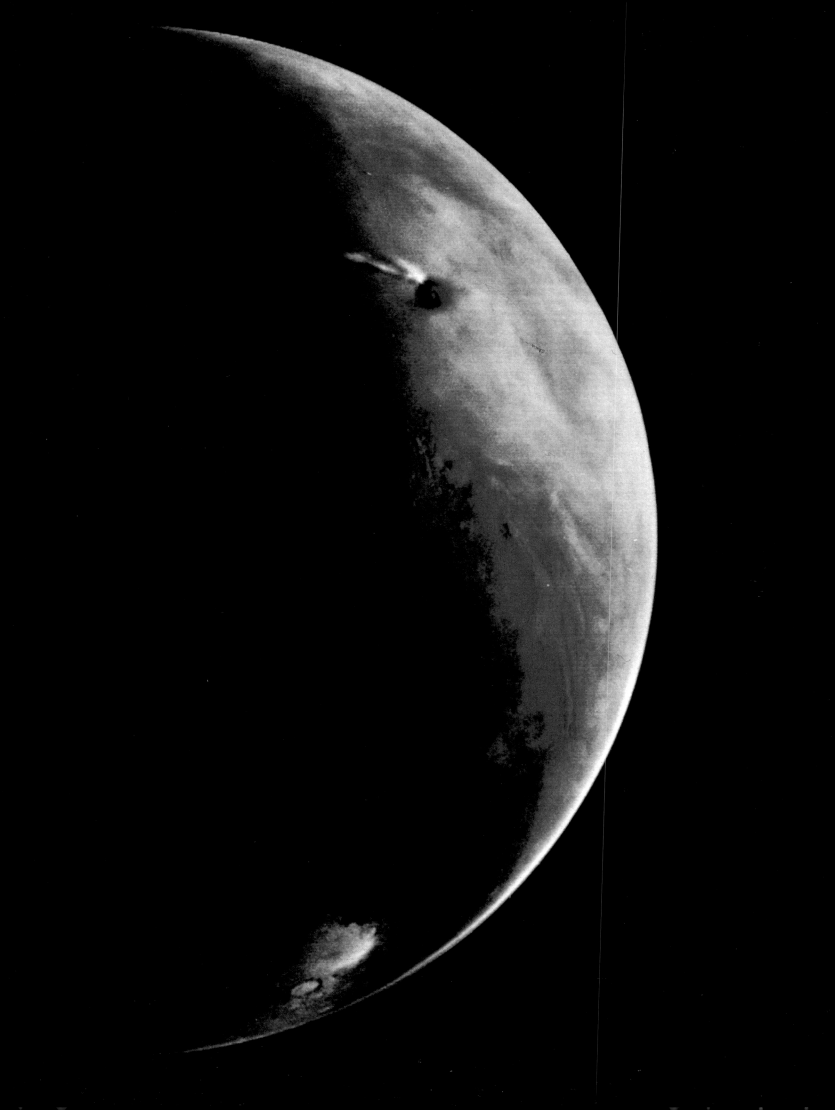

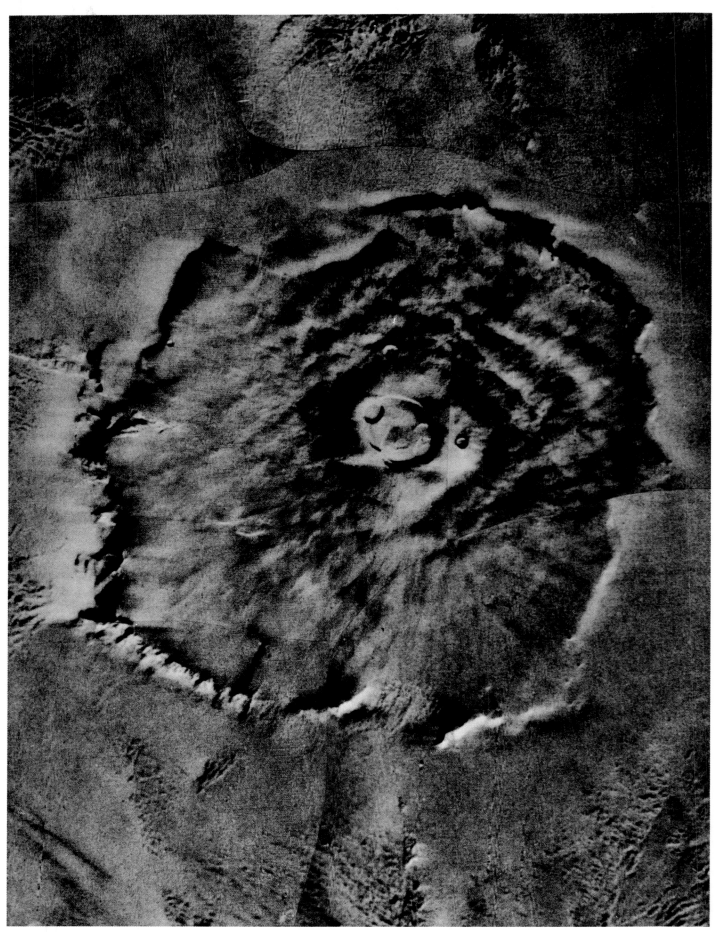

Above: Nix Olympica is the largest known volcano in the solar system. Three hundred miles across at the base and 14 miles high, it is twice the size of the island of Hawaii, the largest vocanic pile on Earth. The cliffs that surround its roughly circular base are more than two miles high. (NASA)

Right: Thin clouds of water ice spill out of the broken terrain in the Noctis Labyrinthus (Labyrinth of Night) on a high Martian plateau at dawn. As the sun warms the atmosphere during the morning, these clouds will evaporate, condensing again at nightfall. (NASA)

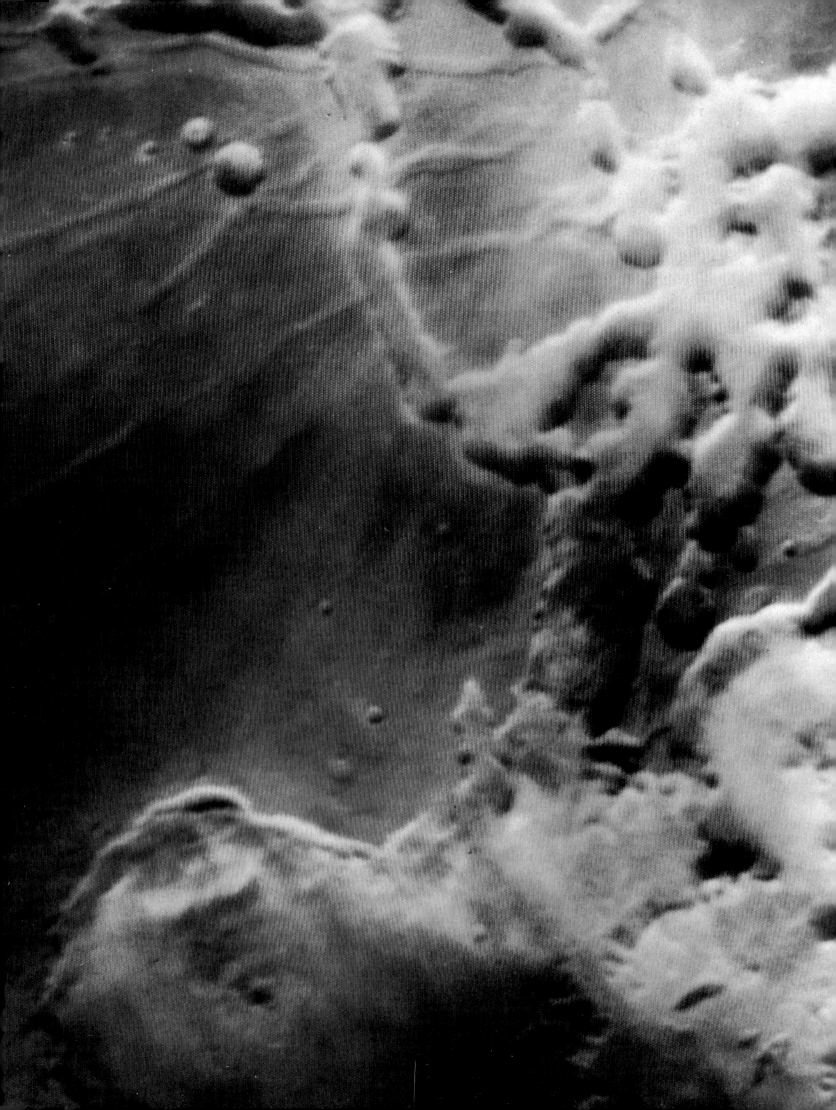

Top: Small dunes surrounding the *Viking I Lander* (the *Lander*'s weather sensor cuts across the picture at the right) closely resemble some found on Earth. They were shaped by wind blowing from the upper left to the lower right in this frame, and they are made of dust; the Marian atmosphere is too thin to transport anything heavier. The large potato-like boulder at the left is 10 feet long and three feet high. (NASA)

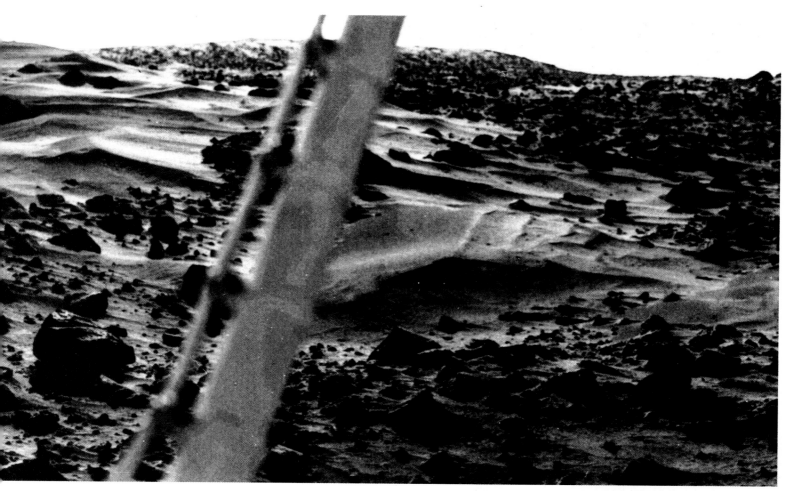

Bottom: Chunks of rock (about two feet across) litter the plain where the *Viking 2 Lander* set down. The landscape is a uniform dull red, getting its color from the oxidation of various minerals, and the afternoon sun shines through a cloudless pink sky. The rocks are of two distinct types: porous (spongelike in appearance) and fine grained. (NASA)

The last light of sunset glows across the Martian surface in a picture made by *Viking I*, scanning slowly from left to right for a 10-minute period. The length of the Martian day is almost exactly the same as Earth's, but the Martian year lasts 687 days. (NASA)

MERCURY

The course of space exploration has been a series of surprises, considering our expectations. We tend to want things to be either fundamentally similar to what we already know or fundamentally different; but our sibling planets, on close inspection, sometimes refuse to follow the patterns we have set for them.

Mercury is an elegant example of how wrong we can be. Until the mid-1960s the planet was thought to be basically similar to the moon, with a landscape dominated by craters and basins and a surface even more hostile to life (if possible) because, like the moon it always presented the same face to the body it orbits — in Mercury's case the witheringly hot sun; one side unremittingly hot and the other side perpetually dark and cold. When infrared observations of Mercury's dark side showed far too much heat for a terrain in permanent night, a thin atmosphere was postulated to carry the heat around from the bright side. (One astronomer even imagined that there was a little tunnel connecting the day side with the night side.) As recently as 1966, researchers were publishing treatises about the Mercurian atmosphere.

We now know that, in fact, Mercury turns its face, rotating at the rate of three times for every two orbits of the sun. This period of rotation was discovered in 1965 by looking for minute differences in radar signals bounced from Earth off the leading edge and the center of the turning planet. Early results suggested that Mercury turned once every 59 days plus or minus 5 days, and Giuseppe Colombo, an Italian expert in celestial mechanics, immediately recognized the ratio of rotation to orbit, correctly predicting Mercury's exact rotation of 58.65 earth days. As for an atmosphere, *Mariner 10* confirmed that Mercury hasn't had one for billions of years.

Such recent confusion about Mercury stems from the difficulty it presents terrestrial observers. Mercury is never farther than 28 angular degrees from the sun in the sky, and it is visible in the night sky only very low on the horizon, just after sunset or before sunrise, through the thickest part of our atmosphere. When *Mariner 10* first swept past the planet in March and April of 1974, its cameras increased the resolution with which we see it by a factor of 5,000.

Mercury is in fact heavily cratered (as had been expected in the moonlike model) but a little smoother, and its surface bears prominent features called lobate scarps — shallow, scalloped cliffs that stretch for hundreds of miles — which we have seen nowhere else. They are vast wrinkles in the surface of a planet whose crust has shrunk around a heavy core as the crust cooled and contracted. The moon never experienced such shrinking, and evidence of early crustal activity on Venus, Earth, and Mars has long since been obliterated by the action of their geologies and atmosphere.

A curious feature of this forbidding planet — one caused by the relation of its rate of rotation to its orbit — is that the sun appears to stop in the sky and move backward for eight days, at noontime, before resuming its slow passage toward evening.

Opposite: A low cliff, called a lobate scarp, runs for hundreds of miles across the surface of Mercury (at the edge of the planet, a little less than halfway up the picture). Mercury's surface is heavily cratered like the moon and is completely airless. (NASA)

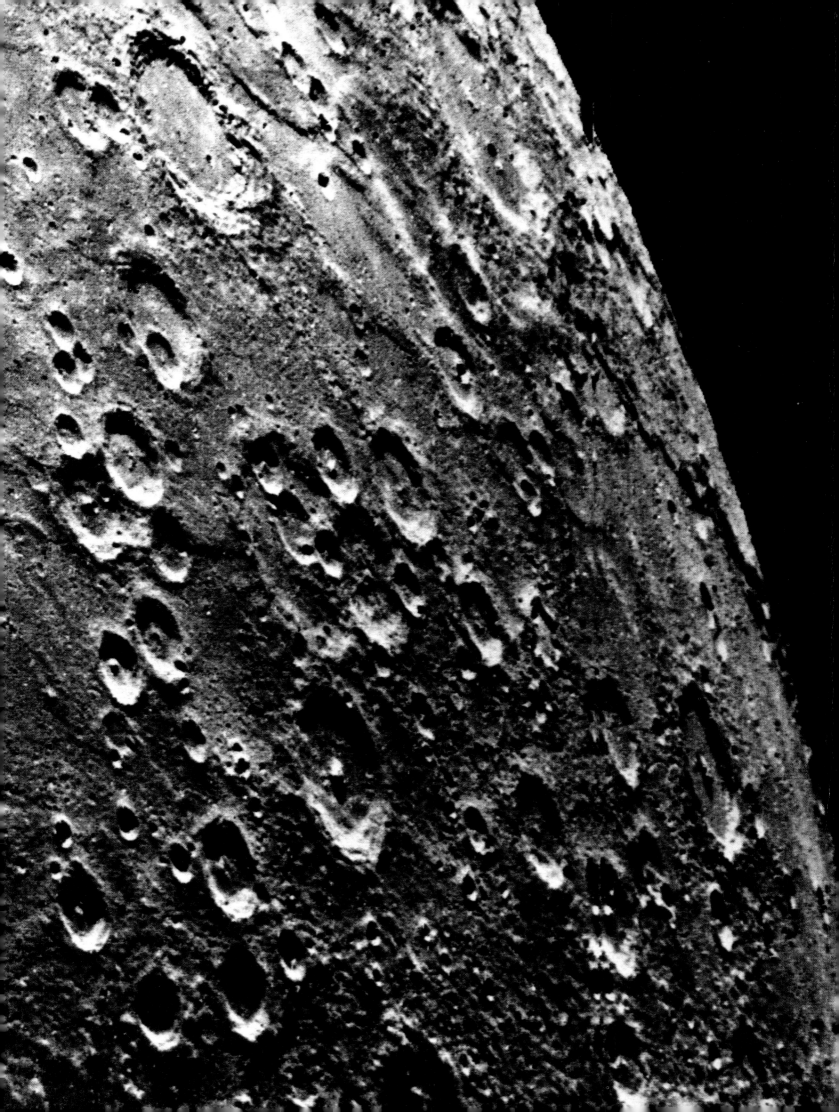

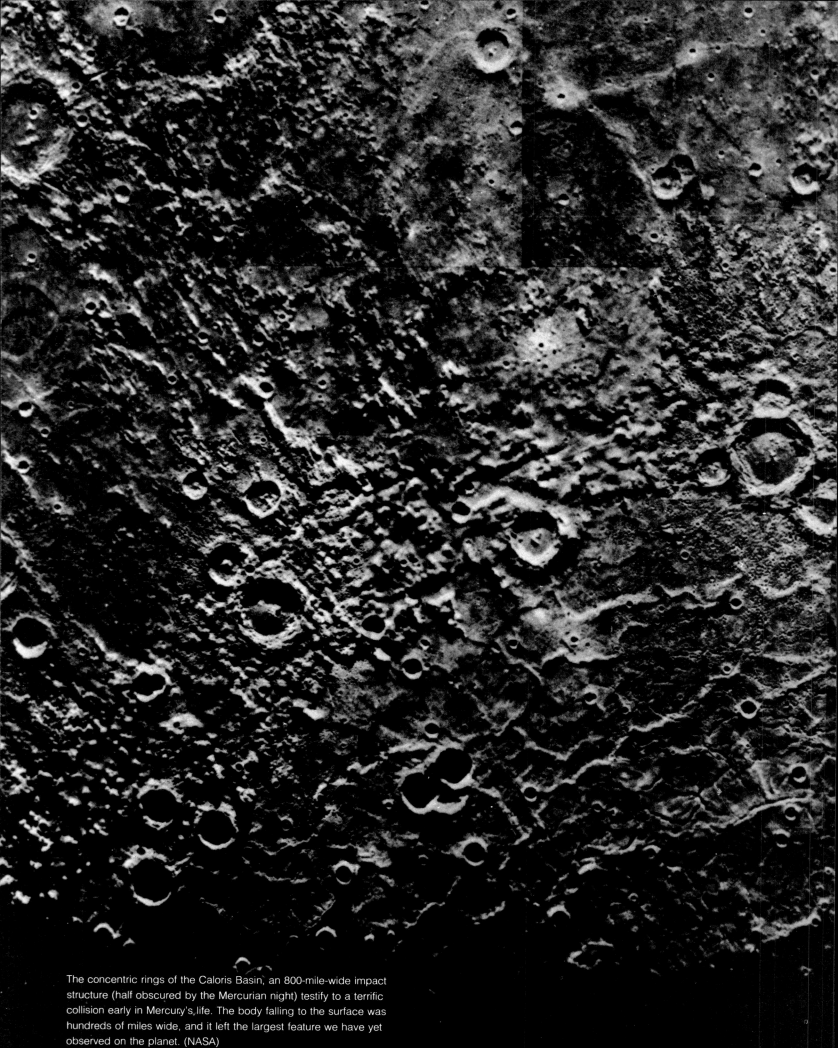

The concentric rings of the Caloris Basin, an 800-mile-wide impact structure (half obscured by the Mercurian night) testify to a terrific collision early in Mercury's life. The body falling to the surface was hundreds of miles wide, and it left the largest feature we have yet observed on the planet. (NASA)

THE SUN

There was a time, not long ago, when we thought of the sun as an essentially undemanding orb. It was like a giant light bulb in the sky, giving off light and heat dependably and doing nothing else. One didn't expect to find solar events influencing terrestrial affairs any more than one expected a light bulb to start moving the furniture.

We currently understand the sun to be both moody and whimsical. In fact, the sun was displaying a certain moodiness at the very moment when sunspots (dark spots on the surface of the sun) were first observed by telescope in the early 1600s. The shocking discovery of imperfections in the sun's surface was immediately followed by an almost century-long period of extraordinary inactivity on the sun's part when, unlike today, sunspots were rarely visible. This activity influenced Earthly affairs in various ways, one of which was to retard our understanding of solar activity.

Perhaps the most telling blow suffered by the "light bulb" model was struck recently by the astronomer John Eddy, who was interested in the sunspot cycle – the (apparently) regular 11-year period during which the number of sunspots rose and fell. Working from tree-ring data (sunspots affect the production of a certain element that is absorbed and retained by the growing tree), Eddy was able to build a solar history of the past few thousand years. His first finding was that the solar circumstances under which we now live (and on which we have built our theories and expectations) are rare in solar history, appearing only 10 percent of the time. Second, the production of sunspots seems like a good measure of the production of solar radiation. When the number of sunspots falls, cooler climatic conditions seem to appear on Earth. (The last sunspot minimum – the one following their discovery in the 1600s – produced a meteorological change that was known as the *little ice age* in Europe.) Third, we are now at an abnormally high peak of sunspot activity and, therefore, are probably in a period of unusually benign temperatures. Finally, what will happen next is unpredictable: the sun may or may not lower its output of radiation at any time, for any length of time.

Our star is a puppeteer with many strings. Earth is connected to it through gravitational and magnetic fields; through the stream of charged particles (called the solar wind) that flows around and past Earth; and through radiation that the sun spills on Earth across the electromagnetic spectrum. All these influences vary (even gravity, causing Earth's orbit to fluctuate slightly and possibly causing some of the ice ages) and they all seem to make a difference: the sun's fitful magnetic field seems to influence the ways in which our atmosphere moves; solar flares are well known to jam radio communications; the sunspot cycle seems to be connected with sustained periods of drought.

It is interesting to note that in our intellectual development as a species we jettisoned an ancient wisdom that conceived of and prayed to the sun as potent and changeable. Our understanding has now come full circle.

Opposite: Our moon fits neatly over the disk of the sun, and periodic eclipses give Earth a view of the solar corona, the white halo extending beyond the sun's obscured surface. The corona is a tremendously hot "atmosphere" (measured in millions of degrees) that plays over a relatively cool surface (measured in thousand of degrees); it is heated by the shock waves of solar radiation emanating from the interior. (NASA)

A vast loop of solar flame twists into space at high velocity. The prominence extends some 250,000 miles above the surface and has a lifetime of only a few hours. The Earth would be little larger than the head of a pin in this ultraviolet picture taken by a special solar telescope on *Skylab*. The light of ionized helium shows the characteristically granular appearance of the surface.(NASA)

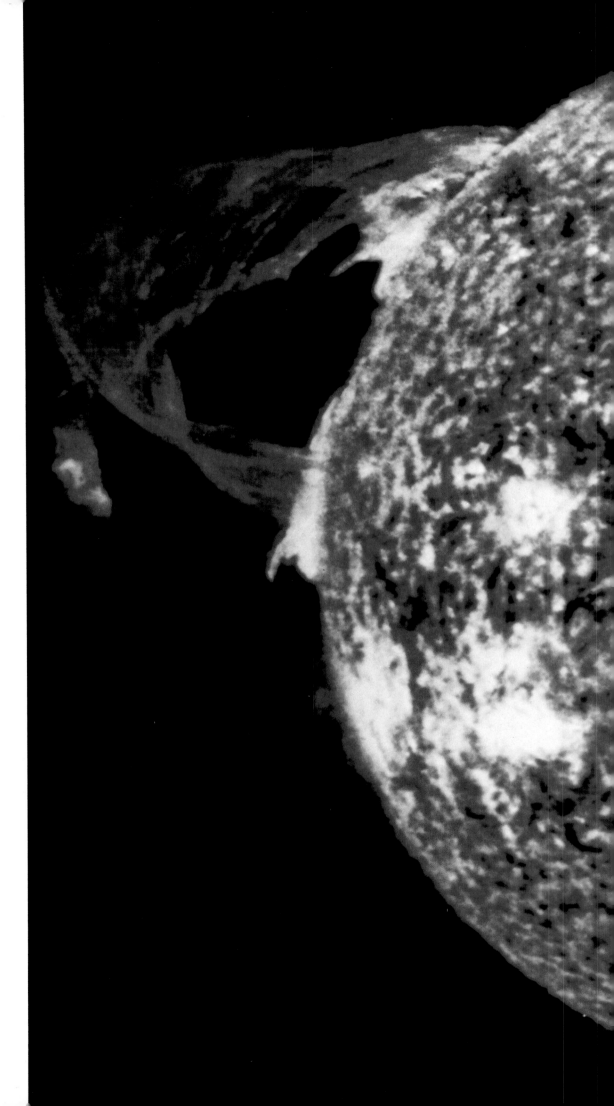

JUPITER

Jupiter is a planet for abstract artists: a kaleidoscope of plumes, streaks, swirls, bands, loops, spots, and patches, running from white through all the browns to bright orange – all in perpetual tumult. Earth aside, it is the liveliest planet we can see.

Its very bulk allows Jupiter to be frothy and changeable. It is large enough to have kept its stock of primordial hydrogen, helium, and other light elements so that the planet is like a thick soup, constantly stirred by Coriolis forces (the same rotational forces that cause our hurricanes to spin.) Jupiter does have some relatively stable features, most spectacular among them being the Great Red Spot that was discovered 300 years ago and has been tracked ever since. It is appropriately large: if Earth were thrown in, it would look like a baseball disappearing into a wishing well.

Some planetologists argue that we do not have nine planets in our solar system but seven, that Jupiter and Saturn, properly considered, are stars. True, they do not burn. But these scientists argue that it is too simple to define "star" by ignition alone; that would be like defining "adult" by specific weight. A star, like an adult, is a pattern of development, a rich complex of characteristics. Ignition and burning are usually part of the pattern, but if they are not, then a body still might be considered a star in other respects.

This school of thought might be illustrated by the observation that the satellite systems accompanying the two gas giants look like planet systems; they have been brought to a pitch of organization that is qualitatively different from the satellite systems of the inner planets. (In fact the inner planets can hardly be said to have systems of any kind, since the only sizable moon in the group is Earth's. Mars has two tiny moons and Venus and Mercury have none.) Jupiter and Saturn, by contrast, each have more than a dozen satellites. Moreover, these are arranged like the planets in the solar system, with the denser, less massive bodies orbiting closest to the planet and the larger, less dense bodies orbiting farther out.

This pattern holds because Jupiter and Saturn are built and behave like stars. They are constructed of hydrogen and helium (not of rock minerals over an iron-nickle core like the inner planets) and are large – large enough, some think, to have started to squeeze their stock of hydrogen together the way stars do when they begin to burn. The two were not quite large enough to achieve ignition, but of sufficient size to drive light elements out to be collected into distantly orbiting satellites. In fact, to this day Jupiter and Saturn generate about twice the energy they receive from the sun.

How close did Jupiter come to ignition? Some believe that the big planet-star would have needed to be about 10 times more massive than it is now to have turned on – but that does not really give an indication as to how close it came. That there are a great many binary star systems in the sky is perhaps a better indication that the chances of our having had two suns were quite good. The delicate balance of our solar system is composed of such near-misses.

Opposite: The giant Jupiter, thought by some to be an unborn star, is 100,000 miles across and contains twice the mass of all the other planets in the solar system combined. It spins once every nine hours, churning its thick atmosphere in the center of a small "solar system" of 13 moons. (NASA)

Above: As Jupiter spins it exerts powerful forces on its atmosphere, causing the regular sequence of belts and zones that are permanent features of the planet. The partial false-color view of Jupiter's north temperate and north polar region (none of which can be seen from Earth) was made from 373,000 miles away by *Pioneer 11* and shows the swirling breakup of belts and zones into continuous rings of hurricane-like storms. Jupiter's north pole is in the night, at the top of the picture. (NASA)

Right: The Great Red Spot is Jupiter's most prominent feature, a centuries-old storm that rages across an area 25,000 miles wide. Jupiter has no surface to interrupt the flow of its atmosphere, so its largest storms drift on for thousands of years before cooling off and dissipating. (NASA)

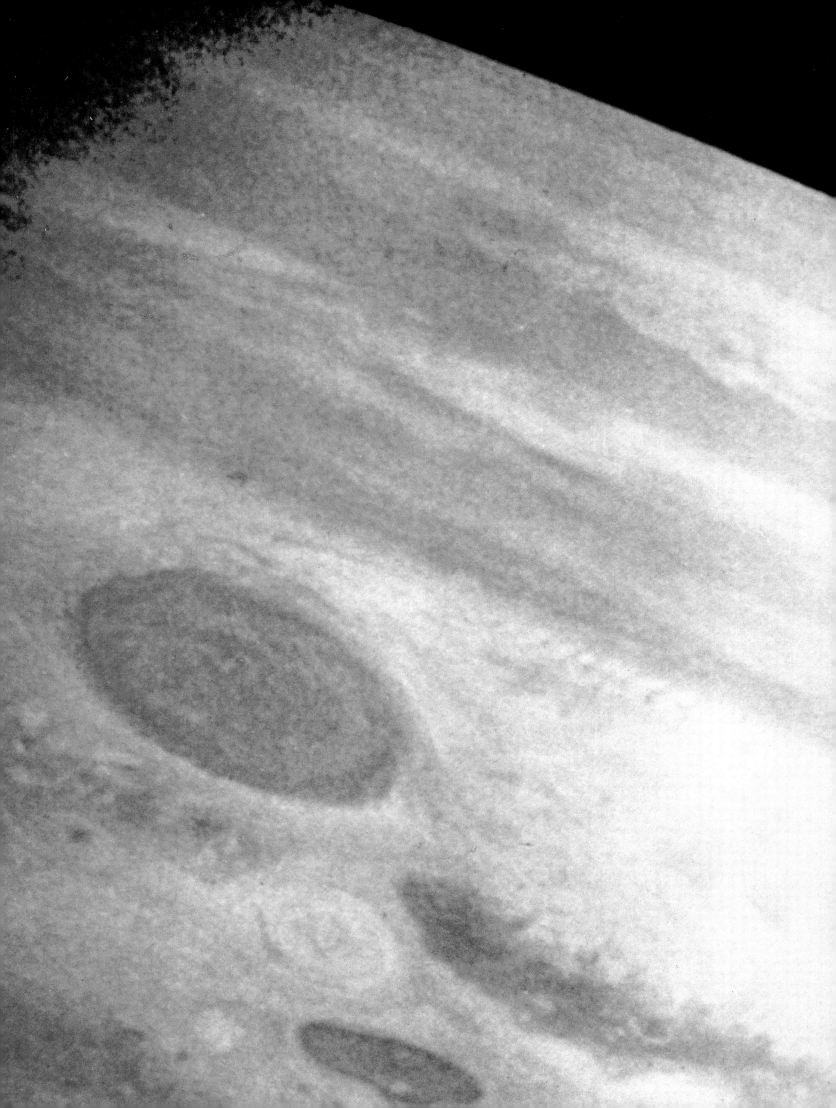

SATURN

On any given night there are probably more hobbyist-astronomers staring at Saturn than at any other planet: only Saturn has the spellbinding rings, the most spectacular ornament in the system.

The rings are composed, for the most part, of rocks of ice condensed out of light elements (hydrogen, helium, oxygen) which once enveloped Saturn. All the planets had these clouds, but in our case we were too close to the sun for the water gases to condense into ice the way they did farther away. After a certain point the sun brightened fiercely and blew all the light gases into the nether reaches of the system. Saturn kept some of its light elements because by that time they had condensed into ice rocks and were more difficult to blow away.

These ice rocks were presumably of all sizes and orbits, but over a period of time they collided with each other so often that they began to look and behave very much alike. As the particles collided frequently, their velocities and motions tended to average out and homogenize. With each collision a little energy was radiated out of the system as heat, and over time the whole group cooled down into a common path. Finally each collision pulverized the rocks a little more. The result was that the ice rocks became smaller (perhaps the size of grapefruits or basketballs), were swept together, and were then flattened out into a snow-white, glittering sweep of ribbon.

The moons of Saturn are less well known than its rings, but they are currently under greater scrutiny by astronomers and, in one case, exo-biologists, the scientists who study the possibility of extraterrestrial life. Iapetus, a small moon orbiting on the fringe of Saturn's system presents the strange appearance of a sphere, half black as asphalt, half white as Earth's polar ice caps. It has been suggested that repeated bombardment of one side of Iapetus may have thrown reflective particles onto the other side, or simply eroded what was once a uniformly reflective body. Another model has the nearby moon Phoebe gradually shedding fine dust which Iapetus (as it travels in the opposite direction) is slowly collecting on its leading hemisphere.

The largest of Saturn's moons, Titan, larger than the planet Mercury, has been an object of intense interest since 1944 when it was observed to have an atmosphere. Recent studies indicate that Titan's atmosphere is surprisingly dense, perhaps as dense as Earth's. Composed mostly of hydrogen and methane, it is thick with clouds.

Titan's surface is probably frozen ammonia and the temperature is 230 degrees below zero Fahrenheit — not conditions that suggest the possibility of life. But add heat to Titan's frigid constitution, and life becomes less inconceivable: if there were volcanic activity on Titan there could be pools of warm water, and given these circumstances Titan would resemble laboratory conditions in which we have observed the beginnings of organic activity. Saturn's huge moon probably does not harbor life, but the possibility makes it an attractive target for exploration.

Opposite: Saturn is tilted on its axis and we see the planet from different angles as it orbits the sun; we get all the views twice for every 30-year round trip. The rings are most prominent when we see the planet from the greatest angle above or below (upper left, lower right), and they disappear when viewed edge-on because they are less than a mile thick. (Lowell Observatory)

Saturn is the lightest planet in the solar system; it would float on water. Composed mostly of hydrogen and helium, it shows faint regular bands of the kind that are characteristic of Jupiter. Four distinct parts of the ring system have been detected; the outermost ring is 170,000 miles in diameter. (Hale Observatories)

THE OUTER PLANETS

Probably the most important fact about Uranus, Neptune, and Pluto, the outermost planets in the solar system, is how far beyond our grasp they are. Even in the age of orbiting observatories, computers, and sophisticated analytical techniques we still know very little about these bodies. The biggest news in astronomy in 1977 was the discovery of Uranus' rings, and they were found by accident, just as Uranus itself was 200 years ago

The discovery of the Uranian rings began with the announcement that Uranus was scheduled to pass between a bright star and Earth. These passages, called occultations, are important events because the size and shape of the planet and its atmospheric structure can be measured by observing the time and rate at which a star dims and blinks off as it is covered by the planet (and then blinks on and brightens as it is uncovered). A number of astronomers outfitted NASA's flying observatory (a large cargo jet) and watched the passage from over the Indian Ocean in order to see the event with as little atmospheric distortion as possible. Because it was not known exactly what moment the occultation would occur, the equipment was turned on early. What it recorded was the star dimming five times before the planet blocked it completely. When the star re-emerged, it dimmed the same number of times at the same rate. To the watching scientists all this added up to the existence of hitherto unsuspected rings.

The tiny scrap of data has been wrung and re-wrung. Specialists tend to the opinion that the Uranian rings are very different from Saturn's: they must be darker or else they would have been de-tected before, and so they probably have more carbon in them than water ice. They seem to be narrower; they probably contain a wider range of particle sizes. The most important difference, from the layman's point of view, is that no one standing on Earth will ever stare awestruck at the rings of Uranus: they are too far away.

Uranus is 15 times as far from the sun as the Earth; Neptune, 17 times; and Pluto is 50 times farther. Even from Uranus' orbit the sun would be no more than a bright point – one star among many. Jupiter and Saturn would be harder to see than Mercury is from Earth. These planets are immensely distant.

As a result we still know very little about the outer planets, and much of what we do know doesn't fit together neatly. Plausible theories about the rotation of the planets do not jibe with plausible theories about their structure. Neptune is farther out than Uranus, but its atmosphere doesn't seem to be any colder. Pluto was discovered in the first place because deviations in Uranus' orbit pointed to the existence of a ninth planet about the size of Uranus or Neptune (about 15 times more massive than Earth), orbiting about 40 or 50 times farther from the sun than Earth's orbit. After 15 years of looking, Pluto was found in about the right place, and a new triumph of the deductive scientific method was announced. But we now know (from occultations) that Pluto couldn't be more than one-tenth the size of Earth's mass!

Opposite: Uranus, the seventh planet, was thought to have only four moons when this picture was taken in 1915, but a fifth moon was discovered 33 years later. The eighth planet, Neptune, and the ninth, Pluto, will exchange places in the solar system for the next several years, as Pluto's eccentric orbit brings it closer than Neptune to the sun. (Lowell Observatory)

STARS

If we take the Big Bang as the Creation, then one has to wonder at how little the Creation actually accomplished. Unlike the Biblical accounts, all the Big Bang did was make hydrogen and a little helium. Hydrogen, the most rudimentary element in the universe, consists of a single proton orbited by a single electron. And the options open to a hydrogen atom are just as limited as one would expect given so stark a nature: it can lose its electron – or not – and it can bind together with one more hydrogen atom to make a molecule of hydrogen gas – or not. All in all, there doesn't seem to be a great deal to work with here.

Obviously something has happened since. We are surrounded by materials and processes that depend on the existence of dozens of radically different elements: elements as light as oxygen, carbon, and nitrogen, and ones more than 10 times heavier, like gold, mercury, and lead. The whole chemistry of life, at least on this planet, is founded on carbon, an element that was not present at the Creation.

What happened after the Big Bang was the development of stars. At some point after the primal flash had subsided, the hydrogen atoms began to collect into clouds and aggregations. The clouds condensed and, under the influence of their gravitational forces, began to contract into stars. (How and why each of these events happened are still unsettled issues.) As the clouds contracted they generated heat and pressure, which mounted steadily to higher and higher levels. The hydrogen molecules were torn apart into their subatomic particles: protons, electrons, and some neutrons. As the heat and pressure climbed, and the particles moved around ever more actively, protons began to collide and fuse, forming new elements.

The first element formed was the next lightest, helium. Two helium nuclei interacting in turn made beryllium; three made carbon. Helium reacted with carbon to make oxygen, then reacted with the oxygen to form neon, and so on. The process continued until iron had been made. Iron nuclei cannot fuse, and one theory holds that the result of generating this indigestible substance is what causes stars to explode, blowing their contents out into space. The interstellar tides sweep these elements back and forth, and (at least in our case) they end up as birds and fish, television sets, and all the rest of life.

Stars are manufacturers, putting together the elements that make our intricate universe possible. Not all stars make all elements; only the brighter, larger, more short-lived stars can make the heavier elements. Our own sun, a yellow dwarf, confines itself to a narrower range of products, but will, as a result, shine longer and more steadily, and perhaps that longevity is a product in itself. In any case, the brighter suns are the ones we see when we look up at night, and the message we can read by their light is that creation continues everywhere.

Opposite: The most detailed view of a star other than our own sun is this computer-enhanced image of Betelgeuse, a red giant star about 500 light-years from Earth. Color is used to show temperature variations on the surface (the hotter portions are orange, the cooler portions blue). If Betelgeuse were set in the center of the solar system, it would fill a volume that extends beyond the orbit of Mars. (Kitt Peak National Observatory)

Following page: The Pleiades are an open star cluster with a few hundred to a thousand members. The brighter stars are surrounded by clouds of gas and dust that look like blue smudges; the circles and crosses on the stars are artifacts of the telescope. This cluster of stars is about 400 light-years away from Earth; our nearest star neighbor is two light-years away. (Hale Observatories)

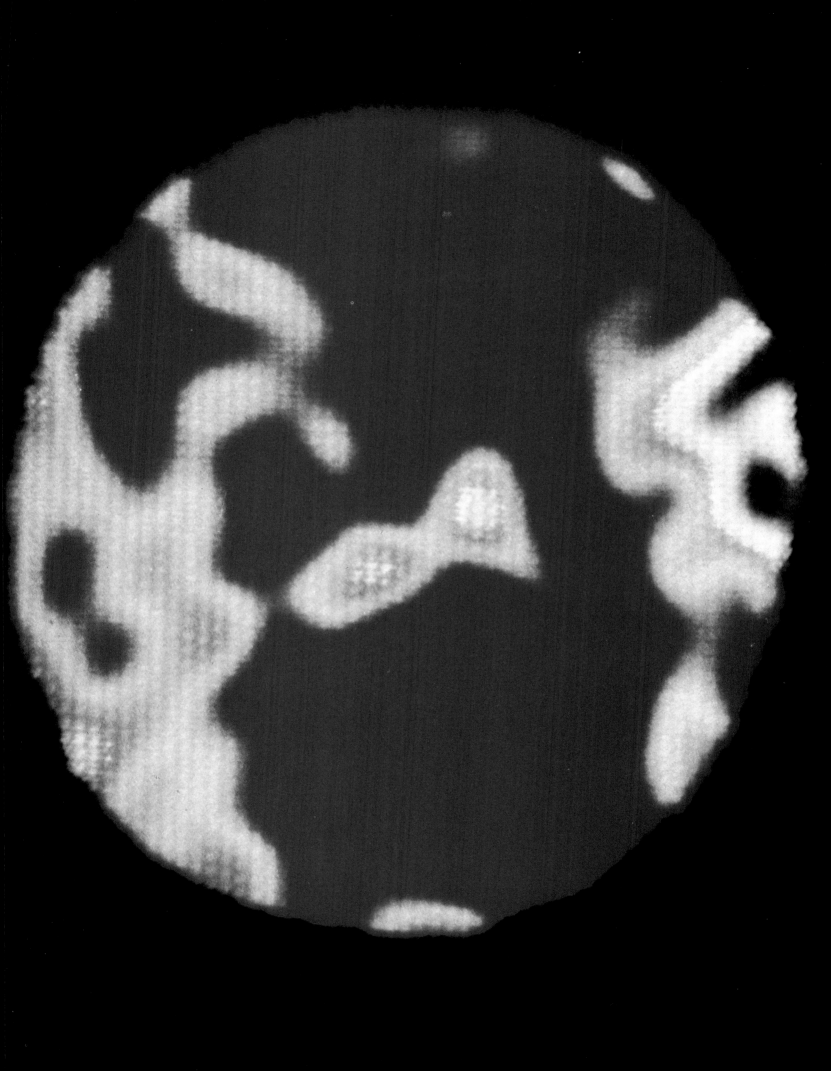

NEBULAE

If we look at our galaxy in terms of a biological metaphor, we might think of the Milky Way as a single organism and the stars as its cells. Like cells, each star has a specific life cycle, and over that life cycle it produces materials that are released into the organism as a whole, giving it its character. In this metaphor the galaxy's circulatory system would be the interstellar medium: the most visible manifestations of which are clouds of dust and gas called nebulae that stream around the galaxy. These clouds recycle the elements made by stars and supernovae and are the raw material from which new star systems are made.

This medium comes from all different stages of a star's life cycle. Some of it is primordial clouds that have never condensed into stars at all; some is from the early stages of star formation, blown off by the fierce solar winds that young stars produce when they ignite. Other materials are "sneezes," whiffs of gas ejected from medium-large stars as they near old age and white-dwarfdom. Still other material is from the interiors of stars, turned inside out and spilled throughout the galaxy by a nova or supernova. From Earth the medium looks like a lacework of interlocked shells; if we could watch it over a long enough period of time, it would look like a network of expanding bubbles, three-dimensional ripples spreading through space.

There is a theory being developed by British and Soviet astronomers that these interstellar clouds make up prebiotic kits, which assemble all materials needed for life independently of the planets. When these kits encounter a planet they are ready to start making a planetary biology: a theoretical process akin to fertilization.

One of the facts on which the theory is based is that dozens of organic molecules have been detected in nebulae. Some are quite complex, and include molecules that can react to make amino acids, which are the building blocks of proteins (which in turn do almost everything in cells). The British-Soviet theory speculates that these molecules might link themselves together on the surfaces of dust grains also held in the clouds and eventually might become so large and complex that they display simple biological functions.

The scientists point to studies made on organic materials extracted from meteorites that show structures similar to geological microfossils. They argue that life appeared on Earth because similar biologically primed meterorites fell into the right conditions here. "It would now seem most likely," one of them writes, "that the transformation of inorganic matter into primitive biological systems is occurring more or less continually in the space between stars."

If this is true, then the circulatory system of the galaxy not only distributes and recycles the chemical elements but also life itself. If so, then we can give up the idea of colonizing other worlds: any other planet that could support us would surely have its own indigenous (and not necessarily alien) life forms already. And all of us everywhere in the galaxy may be descended from the same source: the black reaches among the stars.

Opposite: Hydrogen, energized by the process of star formation, glows in a ring-shaped cloud more than 50 light-years across. (Hale Observatories)

Following page: The Veil Nebula (of which this is only a portion) is the 50,000-year-old remnant of a supernova, the contents of a star blown into interstellar space. (Hale Observatories)

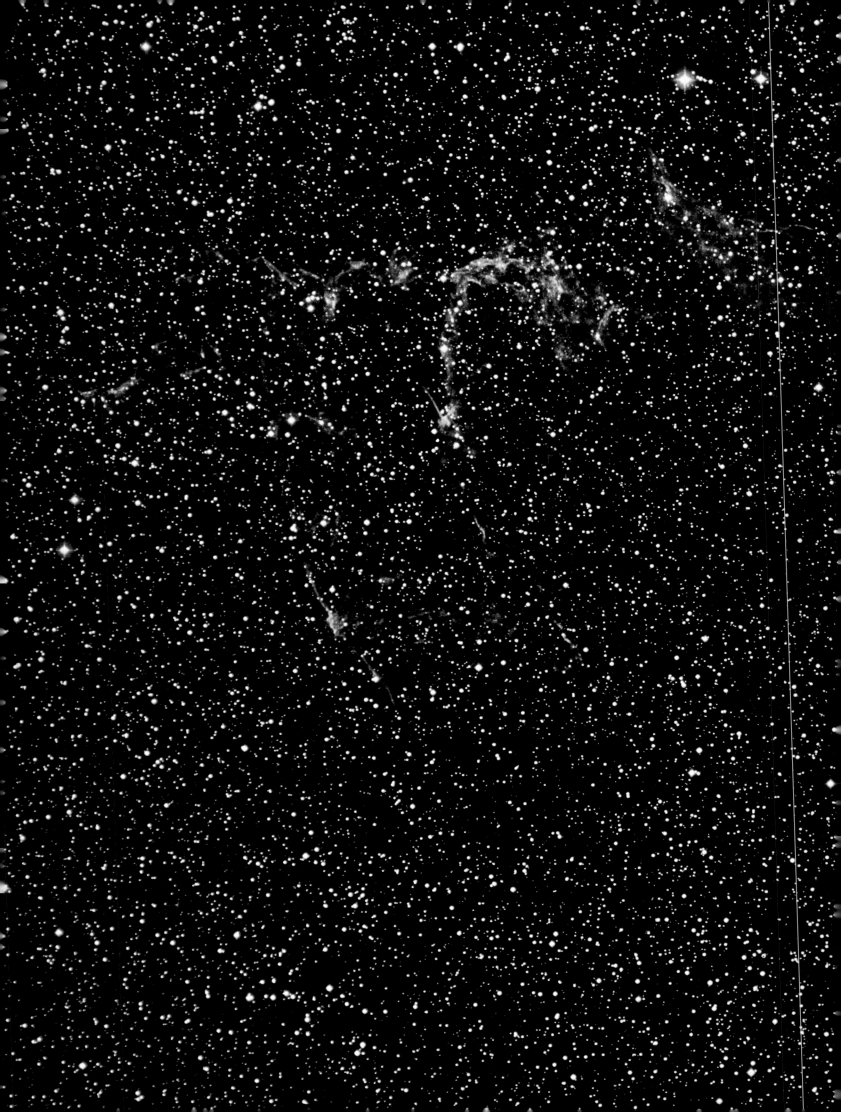

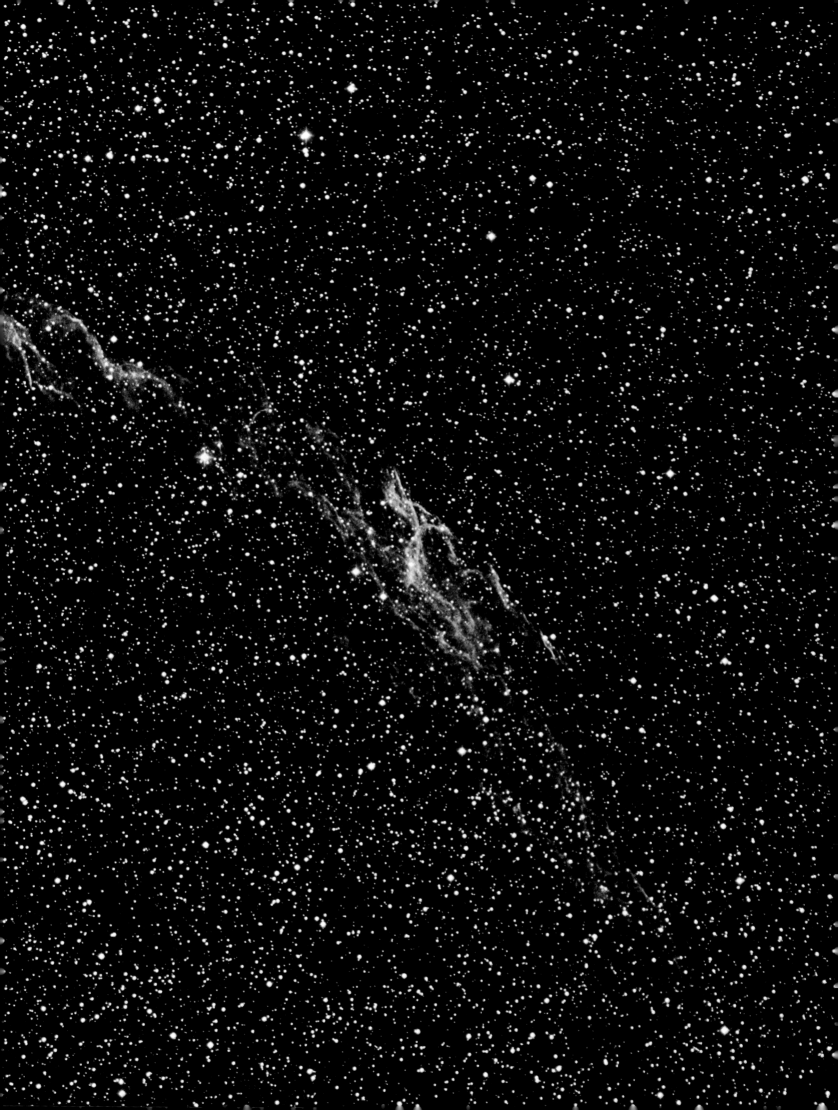

SUPERNOVAE

Events in deep space have influenced terrestrial affairs over and over again – and sometimes decisively. The most common agent of these changes has been the novae: sudden, explosive increases in the amount of radiation being generated by a star, with supernovae being the grandest explosions of all. The invention of writing and arithmetic is sometimes credited to a supernova. The early Sumerians and Egyptians both have myths in which a specific star-god taught mankind to write and count. This god's site in the sky is where a very bright supernova happened around 8,000 years ago. The effort to explain and study the nova is thought to have provoked the development of writing and arithmetic.

Supernovae don't happen often: we have not had one in the Milky Way for 400 years. (By contrast, there are dozens of nonsupernovae in our galaxy every year.) Supernovae are rare because they occur only in a galaxy's largest stars – stars with about 10 times the mass of our sun – and there are not many such stars. Only large stars have enough mass to make iron, and it is iron (astronomers think) that starts the process.

When a large star begins to burn out, its iron core begins to collapse and initiates other energy-draining reactions. The entire inside of the star contracts violently. The masses involved in this contraction are so enormous that the atoms in the star cannot withstand them. Electrons and protons are mashed together, and this creates an enormous burst of energy.

The star layers that were falling in are suddenly blown away entirely. The first layer expands into the neighborhood of the star, sweeping it clean of debris. The pressures of doing so slow it down, and perhaps even force it back toward the supernova. But then the next shell behind, expanding unchecked into the cleared area, hits the first. The impact is as though Jehovah clapped his hands: a tremendous shock wave is produced that sweeps out, shattering all the worlds within dozens of light-years and creating a fireball as bright as 100 billion suns.

The shock wave destroys and it creates. This supernova "sonic boom" is thought to create all the elements heavier than iron (such as mercury, lead, or gold). As it ripples outward, it compresses interstellar gas clouds to the point where their internal forces can start the evolution of new star systems. But the supernova does destroy. Astronomers have calculated that a supernova ought to have occurred within 100 light-years of the sun every 60 million years or so. It seems likely that anything that close to a supernova would have serious problems of some sort. (For one thing, the ozone layer, which protects us against biologically lethal radiation, would be completely destroyed.) And, in fact, the rates at which species have become extinct on Earth have peaked about every 60 million years.

Perhaps the most provocative point about supernovae is that because of them, any decision to confine human civilization to a single solar system is to doom it to destruction by an event that is absolutely certain to occur nearby sooner or later.

Opposite: The top picture shows a distant galaxy before one of the stars on its fringe blows up in a supernova; in the bottom picture the ruptured star rivals the entire galaxy, shining hundreds of millions of times brighter than it did a few years before. (Hale Observatories)

GALAXIES

The discovery of galaxies – more accurately, that the Milky Way is not the only such association of its kind but only one of billions of such aggregations of stars – must rank as one of the greatest discoveries of all time.

Before World War I almost all astronomers believed that the Milky Way was the only significant organization of stars in the universe. Certainly the numbers involved were already too large to make the Milky Way an easy object to understand. It holds, we now know, about 100 billion suns in a volume that takes light 100,000 years to cross. (A hundred thousand years ago modern man – homo sapiens – hadn't yet appeared.)

But even while astronomers were wrestling with the problems of thinking about phenomena on such a scale, during the 1920s a cold, arrogant, ex-boxer named Edwin Hubble discovered a new way to determine the distance between star and Earth, and with this discovery he found that numbers like 100 billion stars were peanuts.

The real story is that the universe is overflowing with galaxies. The latest estimate is that there are about as many galaxies in the universe as there are stars in the Milky Way, contained in a volume that takes light about 17 billion years to cross. Numbers as large as these are, of course, essentially meaningless. What they say, in part, is that certain aspects of the universe may never be understood even when they can be measured.

Hubble's discovery immediately triggered another: that the farther out one looked in any direction the faster these galaxies seemed to be fleeing from the Milky Way! The observation had been predicted by Einstein's general theory of relativity in 1916 (but not by Einstein, who hadn't been able to bring himself to contradict the opinions of the established cosmologists; he had assumed his theory was in error in this regard). The theory of relativity said, in part, that the universe was constantly expanding and therefore everything in it should be continually moving away from everything else except for those objects that are gravitationally linked, like the bodies in the solar system or the stars in a galaxy.

The idea that the universe is expanding immediately suggests one theory of its origin: that the universe first was in an unexpanded state and then blew up. The same theory offers a way to get a peek at this event. The farther apart two galaxies are, the longer it will take for light to travel between them; when light arrives at galaxy A from galaxy B, the information it carries will be about the conditions on B when the light started its journey. If we want to look at a galaxy as it was a million years ago, we should try to find one a million light-years away; if we want to see how things were 10 million years ago, we should try to find something 10 million light-years off, and so forth.

The universe is thought to be about 17 billion years old, and if we find something that far away in space-time (the theory of relativity says), we will be looking straight into the heart of the Big Bang itself. The rule of the route back to the origin is to look up and out – to the oldest light coming from the youngest galaxies.

Opposite: One hundred billion stars move together through space in a spiral galaxy about 25 million light-years distant. The Milky Way would look the same if seen from this angle; if this were the Milky Way, our sun would be located in one of the curving arms, about two inches from the center of the picture. (Hale Observatories)

Above: The Sombrero galaxy, named for its resemblance to the hat, has a thick band of dust around the edge of its outer structure. One hundred thousand light-years across, it is 14,000,000 light-years from Earth. (Hale Observatories)

Preceding page: The galaxy in Andromeda is the Milky Way's only large neighbor in a group of about 20 galaxies. Its traditional name is the Andromeda Nebula, from a time before it was understood as a system outside the Milky Way; it is about 1 million light-years from Earth. (Hale Observatories)

Right: While two galaxies pass near enough to interact, the larger one exerts enough gravitational pull to mix their materials. The image is processed in these colors to increase the differentiation of areas that would otherwise be indistinguishable shades of gray. (Kitt Peak National Observatory)

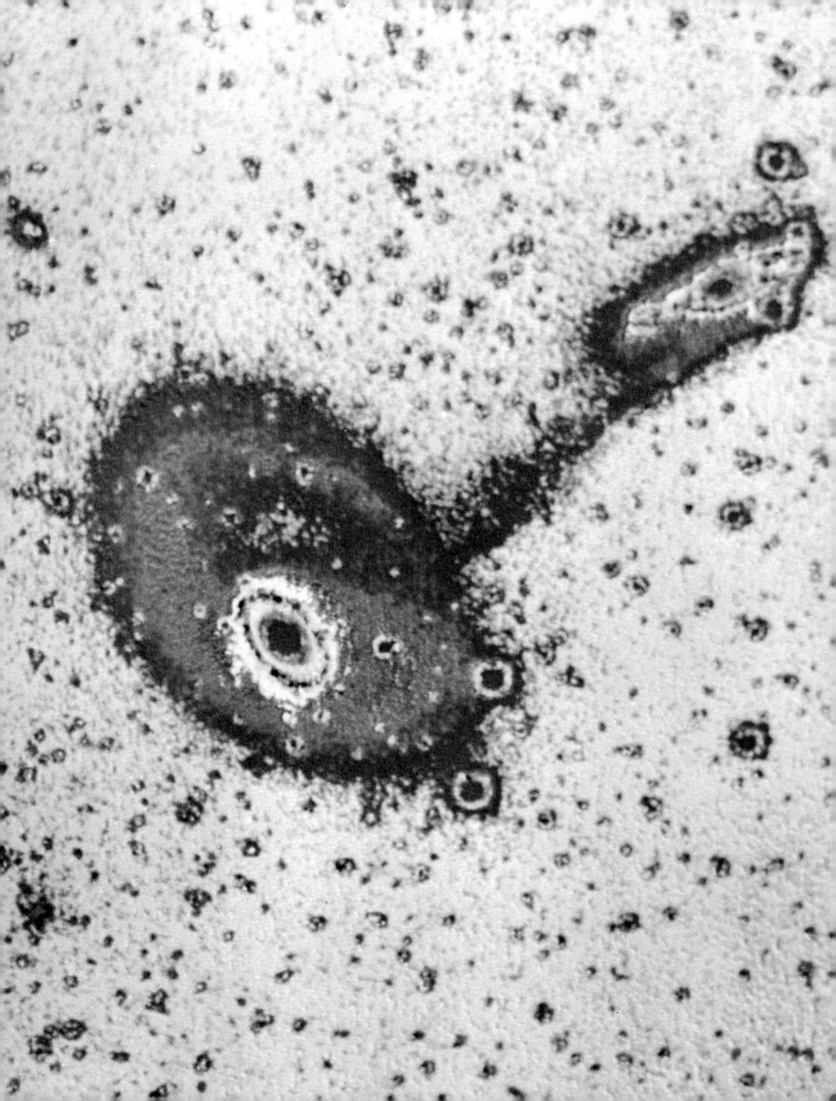

GALAXY CLUSTERS

Scanning the hierarchy of size in the universe, from planet to solar system to galaxy, one keeps expecting to encounter some object large enough to dominate all external influences – a cosmic individual in complete control of its fate. But not even our galaxy, with its 100 billion suns, is independent and self-sufficient (acting according to forces that flow out of its own nature and not those imposed from without).

There are about two dozen galaxies associated with the Milky Way in a small galaxy cluster called the "local group." The only one larger than the Milky Way, Andromeda, with its 300 billion stars, travels on the other side of the cluster – about a million light-years away. Andromeda and the Milky Way together make up most of the local group's mass (70 percent); the other members of this cluster are small fish by comparison.

A great many galaxies exist in much larger clusters, where the society of other galaxies can affect them as intimately as the pressures of urban life affect city dwellers. One clue to this fact came to light when astronomers began to find a pattern of high energy production in large, rich galaxy clusters. The nearest of these to Earth is a cluster with about 1,000 galaxies in the constellation Virgo. In its center is an especially luminous galaxy referred to as M87. This has turned out to be quite a remarkable object. For one thing, X-ray detectors carried by the *Uhuru* satellite found that M87 was enveloped by an X-ray-emitting cloud nearly a million light-years across. (A million light-years is 10 times the diameter of the Milky Way, about the width of our entire cluster.) This is a substantial amount of material for one galaxy to be holding on to. Indeed, calculations demonstrate that to do so, M87 needs to have a mass several hundred times larger than the Milky Way. These super-giant galaxies, which is what astronomers call them, have been found in a number of rich galaxy clusters. They are thought to have evolved through the interaction of galaxies moving near enough to exert gravitational forces on each other that slowed them down, changed their orbital paths and eventually melded them into single super-giant galaxies.

Surely the super-giants are the cosmic individuals mentioned earlier. Can there be anything so large that it could affect the development of a super-giant as it gradually consumes the large galaxies of a rich cluster? Probably not, unless . . .

There are certain astronomers who point out that the nearby clusters all seem to be oriented on the same plane, as though they had all been affected by a common force. And astronomers say that if you look at the galaxies and clusters around our own cluster, and do so for a region about 100 million light-years across, you begin to notice a thickening, a great circle, that looks a little like an equator. If so it could only be the equator of a gravitational field that operates on the level of associations of clusters – a cluster of clusters. Perhaps such aggregations do not exist (one rather hopes not); but if they do, the forces they could exert might make even the super-giant galaxies seem small.

Opposite: Stephan's Quintet are five galaxies in remarkable proximity to each other, although some astronomers argue that, in fact, only four are interacting and that a fifth galaxy is much nearer to Earth. Black on a white background is a traditional way to process astronomical photographs for research purposes. (Kitt Peak National Observatory)

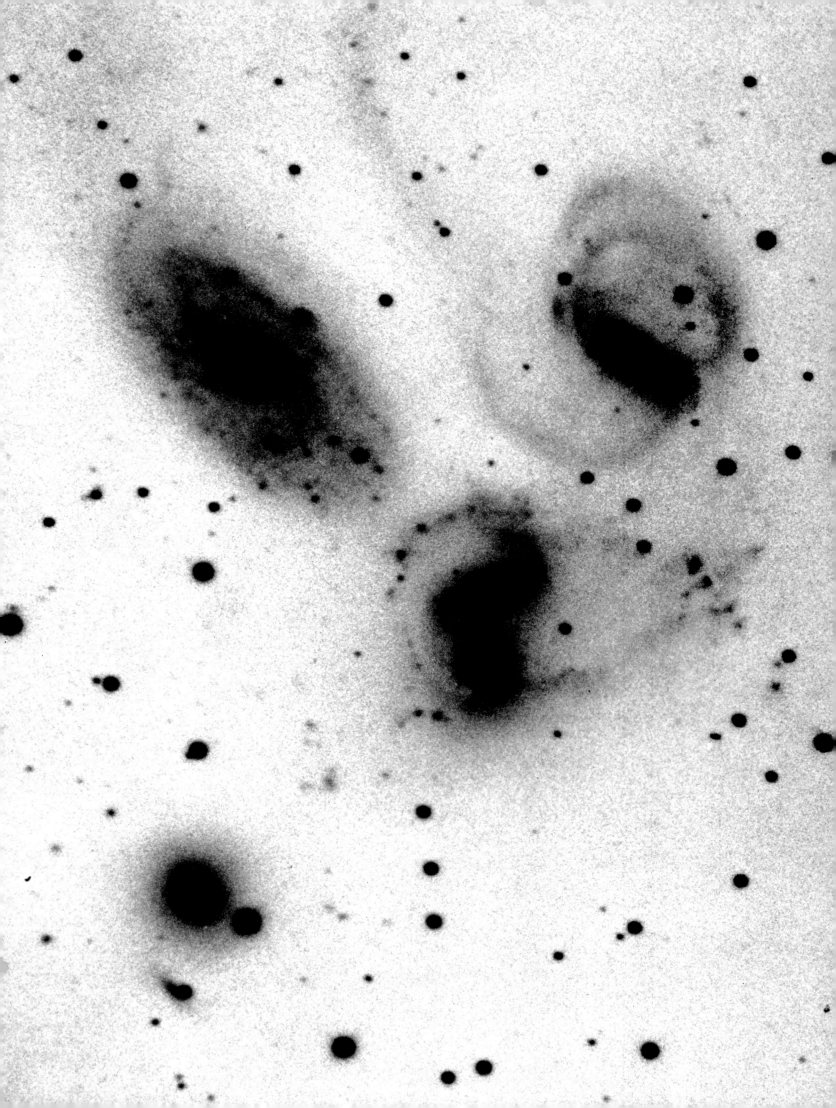

Thousands of galaxies, each with
billions of stars, move together in a
cluster in the constellation Her-
cules. When the light that produced this image started toward
us more than 600 million years
ago, the highest lifeforms on Earth
were invertebrate sea creatures.
(Hale Observatories)

BLACK HOLES

Many scientific theories are not so much arguments about the way nature *is* but the way it *might* be; these theories can make us less, rather than more, certain about what is true. "Does this mean so-and-so is real?" one might ask the author of a theoretical paper. "Oh, no" is the likely response, "it only means that you can put forth a good argument that it *might* be true."

The idea of black holes arose in the early 1960s as a purely logical deduction from the general theory of relativity. When an object becomes so dense that escaping from it requires velocities greater than the speed of light, then nothing can escape, not even light. Very high density is required to achieve this kind of gravitational pull — Earth would have to be squeezed to the size of a cherry to accomplish this.

During the 1960s theorists played around with the idea of black holes. At the center of the black hole, calculations showed that matter would be squeezed out of existence altogether! But what did that mean? Perhaps that matter was being sucked into another universe? Or another place in the universe? Or another time? During this period of speculation, no actual black hole had been found.

Then, in 1970, the *Uhuru* X-ray satellite found X rays coming from a neighboring star in the constellation Cygnus. Ordinarily stars don't generate X rays; so scientists looked a little closer. Certain cycles were found in the spectrum of radiation emitted from this star, which indicated that the star was orbiting around a body with a mass about 10 times that of our sun. If stars don't make X rays, then the X rays *Uhuru* saw must have been coming from this dark companion. The X rays themselves came in patterns that suggested they originated from a quite small body. What is there that is both that small and that massive? If all the speculation about black holes had not been accumulating for 10 years, astronomers might have been stumped for an answer; as it was, they had the right one already on the shelf. The announcement was made: a real black hole had been discovered.

Now that a real black hole had been found, the cycle of speculation began again. Suppose that mini black holes are being made right now inside neutron stars (neutron stars are also very high-density objects, only not quite as dense as black holes). A star with a black hole for a core opens up several possibilities. The black hole might even eat the rest of the star in an instant, releasing so much energy that it would look like a supernova outburst.

Or imagine that there are super-massive black holes measuring up to billions of solar masses. Such an enormous gravitational field would suck in stars like atoms of hydrogen. Perhaps super-massive black holes are what lie behind some of the mysteriously energetic spots in the sky.

Are these speculations true? Probably not. On the record, one would have to say that the odds in favor of any scientific theory lasting for more than a few years are very poor. But enduring truth is really just a by-product of the real sport, which is trying to be the first one to guess what the universe is going to do next.

Opposite: Recent studies suggest that the center of the giant radio galaxy M-87 may be a black hole 5 billion times more massive than the sun. M-87 is surrounded by an X-ray emitting cloud 1 million light-years across. (Kitt Peak National Observatory)

QUASARS

n the early 1960s astronomers began to discover objects in the sky with radiation patterns that were utter anomalies, completely unlike the broadcast patterns of any other celestial objects. They looked somewhat like stars; so astronomers named them quasi-stellar objects – quasars. After about two years of puzzling over these patterns, two California astrophysicists had a profound insight: all the observations made sense if one assumed that these quasars were receding at speeds approaching that of light. But if they were receding at such speeds, they would necessarily be very far away – and to be so distant and yet be bright enough to look like stars would mean that they were putting out more energy than our entire galaxy.

As more and more quasars were found, certain patterns emerged. Whatever these objects were, it seemed clear that they had been a feature of the early universe.

There are very few quasars whose light has been traveling for "only" 2 or 3 billion light-years. The number of quasars grows rapidly as we look farther into space and back into time – until we see objects whose light has traveled for 15 or 16 billion light-years, about 90 percent of the age of the universe. Past this point no quasars have been found.

All around the world theories have bloomed in astrophysical journals. Some say they might be white holes – black holes reversed in which material was expanding *into* existence instead of being squeezed out of it. Others propose dense clusters of stars, members of which would be supernovaeing at just the right rate to account for the observed energies. Others argue that the distance observations must be wrong, that quasars are really not so far away, and therefore not so bright. Still others propose that quasars are supermassive magnetized rotators, the sites of matter/antimatter interactions.

The majority view, at the moment, is that quasars throw off so much energy because they have black holes in their hearts. The brightest objects in the sky are to be understood by the proposition that they are built around the blackest, least visible bodies in the universe. These black holes are imagined to have a mass several billion times that of our sun, powerful enough to raise a "black tide" on any star that orbited nearby, that is, a tide powerful enough to pull the star apart. The debris from destroyed stars and other sources would, it is imagined, form an accretion disk of gases that spiral into the hole. As these gases are pulled down into the hole they accelerate to such high levels of energy that they could produce the bright lights known as quasars.

It is certain that the story of these cosmic candles has only begun to unfold. So far, it is a tale of contrasts: the darkest object of all making the brightest; the bodies farthest from us throwing broad hints about what lies in the center of the Milky Way, practically in the next room, relatively speaking. All that can be predicted is that we will continue to find sights of which we could never have dreamed, that we will try to match the stride of our imagination with what we see, that we will fail, and, in failing, not fail.

Opposite: The quasar, or quasi-stellar object, called 3C 273 is 100,000 to 1 million times brighter than the Milky Way. If present determinations of distance (about 2 billion light-years) are correct, the unexplained jet shooting out from the quasar's center may be 150,000 light-years long. (Kitt Peak National Observatory)